Nanomaterials for Energy Applications 2013

Topical Conference at the 2013 AIChE Annual Meeting: Global Challenges for Engineering a Sustainable Future

San Francisco, California, USA
3-8 November 2013

ISBN: 978-1-63439-025-5

Printed from e-media with permission by:

Curran Associates, Inc.
57 Morehouse Lane
Red Hook, NY 12571

Printed by Curran Associates, Inc. (2014)

AIChE
3 Park Avenue
New York, NY 10016-5991

Phone: (203) 702-7660
Fax: (203) 775-5177

www.aiche.org

Additional copies of this publication are available from:

Curran Associates, Inc.
57 Morehouse Lane
Red Hook, NY 12571 USA
Phone: 845-758-0400
Fax: 845-758-2634
Email: curran@proceedings.com
Web: www.proceedings.com

TABLE OF CONTENTS

(65a) Templated Ordered Mesoporou Materials for Supercapacitors and Lithium Ion Batteries

Lee, J., Pohang University of Science and Engineering

Originally presented on: 11/4/2013 8:30:00 - 9:00:00

For high performance supercapacitor and Li-ion battery electrodes, various mesoporous metal oxide materials have been synthesized by employing lab-synthesized block copolymers as a structure-directing agent or ordered mesoporous silica materials as hard templates. An ordered mesoporous WO_{3-X} material with high electrical conductivity was employed for use as a supercapacitor electrode in 2.0 M H_2SO_4. The electrode made of mesoporous WO_{3-X} exhibited a high rate capability and an excellent capacitance (366 Fcm^{-2}, 639 Fcm^{-3}), which were attributed to the large ordered mesopores, high electrical conductivity, and high material density. Fe_3O_4 nanocrystals confined in mesocellular carbon foam (MSU-F-C) are synthesized by 'Host-Guest' approach and tested as anode for lithium-ion batteries (LIBs). 13-27 nm magnetite Fe3O4 nanocrystals are formed inside ~30 nm pores of mesocellular carbon matrix. Fe_3O_4/MSU-F-C (45 wt% Fe_3O_4) exhibited 1007 mA h g^{-1} of first charge (Li de-insertion) capacity at 0.1 A g^{-1} and 111% of capacity retention at 150th cycle, and retains 37% capacity at 7 A g^{-1} (~7 C rate). For the first time, we synthesized mesostructured LTO/carbon composite (Lithium titanate/carbon) with large (> 15 nm) and uniform pores for high performance LIB anodes via a block copolymer assembly. Sn-mesostructured carbon nanocomposites have been synthesized by simple self-assembly of block copolymers with Tin and carbon precursors.

(65b) Reduced Graphene Oxide Wrapped FeS Nanocomposite for Lithium-Ion Battery Anode With Improved Performance

Fei, L., New Mexico State University
Luo, H., New Mexico State University
Originally presented on: 11/4/2013 9:00:00 - 9:15:00

The fast growing business of portable electronics (e.g. computers, mobile phones, cameras) has stimulated the rapid development of lithium-ion batteries (LIBs) due to their high energy density, flexible and lightweight design, and longer lifespan than conventional battery technologies. Graphite is the most popular commercial anode. However, its low theoretical capacity (372 mAh/g) renders it hard to satisfy the increasing market demands. Recently, metal sulfides with high theoretical capacity are widely studied as another promising class of anode materials with excellent electrochemical performance.

Here, we report FeS based nanocomposite as anode for lithium-ion batteries. FeS nanoparticles wrapped in reduced graphene oxide (RGO) are produced via a facile direct-precipitation approach. The resulting nanocomposite FeS@RGO structure has better lithium ion storage properties exceeding those of FeS prepared without RGO sheets. The nanocomposite delivered an initial discharge capacity of 1357 mAh/g and a charge capacity of 1116 mAh/g, giving the first cycle coulombic efficiency of 82 %. The composite shows a very stable cyclic performance with its capacity remaining as high as 978 mAh/g even after 40 cycles at a current density of 100 mA/g. This value is almost three times the capacity of commercial graphite anode, and is comparable to those reported in literature. At current density of 300 mA/g, the composite delivered discharge capacity of 927, 690, and 618 mAh/g in the first, 10th, and 30th discharge cycles, respectively; while FeS suffered rapid capacity decline, only delivering 885, 518, and 382 mAh/g in the first, 10th, and 30th discharge cycles. Electrochemical impedance demonstrate that FeS@RGO has a lower electrochemical resistance than that of FeS. The enhanced electrochemical performance is attributed to the robust sheet-wrapped structure and the synergetic effects between FeS and RGO sheets, such as increased conductivity, shortened lithium ion diffusion path, and the effective prevention of polysulfide dissolution.

(65c) Alkanethiol-Passivated Germanium Nanowires As High-Performance Anode Materials for Lithium-Ion Batteries

Tuan, H. Y., National Tsing Hua University
Yuan, F. W., National Tsing Hua University
Originally presented on: 11/4/2013 9:15:00 - 9:30:00

Rechargeable lithium-ion batteries have been widely used in portable electronic devices such as cell phones, laptop computers, and digital cameras, and have been considered as an electrochemical energy storage system for powering hybrid electric vehicles (HEV) and zero-emission vehicles (ZEV). It is necessary to develop lithium-ion batteries having a higher energy density and a higher power density to meet the increasing energy demand. Germanium (Ge) has emerged as a promising anode materials due to its high theoretical capacity of 1384 mAh/g - around four times greater than commercial graphite anode (372 mAh/g). However, the practical usage of Ge as an anode material is hindered by dramatic volume changes (~300%) during the insertion/extraction process of lithium ions, causing crack and pulverization, and loss of electrode contact.

Ge nanowires can effectively accommodate the volume changes, tolerate relaxed mechanical strain, provide channels for efficient electron transport, and alleviate pulverization. Although direct incorporation of Ge nanowires into slurry coating process shows improved electrochemical performance relative to microparticles, the anode performance decayed rapidly within a few tens of cycles. In this talk, we demonstrate that dodecanethiol monolayer passivation on Ge nanowire surface can significantly enhance their anode performance in lithium-ion batteries. The dodecanethiol-passivated Ge nanowires exhibit an excellent electrochemical performance with a reversible specific capacity of 1130 mAh/g at 0.1 C rate over 100 cycles, which is above 80% of the Ge theoretical capacity. The functionalized Ge nanowires show high-rate capability having charge and discharge capacities of ~555 mAh/g at rates as high as 11 C. An aluminum pouch type lithium cell using a $LiFePO_4$ cathode was also assembled to provide larger current (~30 mA) for uses on light-emitting-diodes (LEDs) and audio devices. Organic monolayer-coated Ge nanowires represent promising Ge-C anodes with very low carbon content (~2-3 wt %) for high capacity, high-rate lithium-ion batteries and are readily compatible with commercial slurry-coating process for cell fabrication.

(65d) A Facile Microwave-Assisted Route to Co(OH)2 and Co3O4 Nanosheet for Li-Ion Battery

Chen, G., New Mexico State University
Luo, H., New Mexico State University
Originally presented on: 11/4/2013 9:30:00 - 9:45:00

Cobalt hydroxide ($Co(OH)_2$), as one of the most important functional inorganic materials, can be used in the field of catalysis and electrochemical electrodes. The previous report revealed that $Co(OH)_2$ is a promising alternative conversion-based anode materials for lithium-ion batteries (LIBs) due to its high capacity. Unfortunately, like other conversion-based anode materials, the implementation in LIBs is greatly hampered by its poor cycling performance. One of the most promising strategies to tackle this obstacle is to make $Co(OH)_2$ into sheet-like structure to improve the structural stability in the electrochemical reactions. Compared with the bulk $Co(OH)_2$, $Co(OH)_2$nanosheet displays obvious advantages due to large surface area, highly oriented growth direction, tunable interlayer space for the de/insertion of lithium ions.

In this work, a facile microwave-assisted synthetic route has been successfully developed for preparing hexagonal $Co(OH)_2$ nanosheets with average width of 2 μm and thickness of 100 nm. $Co(OH)_2$ can further convert to porous Co_3O_4 nanosheets *via* thermal decomposition. Their phases, structures and morphologies were investigated by X-ray diffraction (XRD), atomic force microscopy (AFM), and transmission electron microscopy (TEM). Both $Co(OH)_2$ and Co_3O_4 nanosheets can serve as potential candidates for anodes of Li-ion battery. The electrochemical study revealed that $Co(OH)_2$ and Co_3O_4 nanosheets delivered a reversible capacity of 600 and 700 mAh g^{-1} after 40 cycles, respectively. Cyclic voltammetry (CV) curves also confirmed the relative stability of the as-synthesized electrode materials. This effective microwave-assisted route may be a promising approach for preparing other transition metal hydroxide/oxides for energy applications.

(65e) Li Adsorption and Intercalation in Graphene, Graphite and Carbon Onions By Reaxff Reactive Force Field

Raju, M., The Pennsylvania State University
Ganesh, P., Oak Ridge National Laboratory
Kent, P. R. C., Oak Ridge National Laboratory
van Duin, A. C. T., Pennsylvania State University
Originally presented on: 11/4/2013 9:45:00 - 10:00:00

A ReaxFF reactive force field has been developed to describe Li/C systems. The ReaxFF force-field parameters for Li-C systems were optimized against a density-functional theory (DFT) (with van der Waals corrections) based training set, containing a collection of data (energies, geometries, charges etc.) related to adsorption and diffusion of Li on graphene and in graphite. ReaxFF reproduces accurately the adsorption energies for Li on hollow, bridge and top sites on pristine graphene, formation energy of stage I compounds in graphite and intralayer Li migration barriers in graphite at separations of 3.35 Å and 3.7 Å. To validate this force field we have compared it against DFT calculations for Li diffusion in the direction perpendicular to the graphene sheet for graphene with single and double-vacancy. We performed grand canonical Monte Carlo simulations to obtain ground states for Li intercalation in graphite and find that ReaxFF describes the in-plane Li ordering and interlayer separations for stage I and II compounds in good agreement with previous DFT studies and experiment. The voltage profile for the Li-graphite system as a function of Li concentration obtained by ReaxFF matches experiment and DFT data. We study Li intercalation in presence of vacancy defects in graphite and find that the presence of vacancy defects increase the ratio of Li/C and shifts the voltage profile downwards, both in proportion to the number of vacancy defects. Li intercalation and diffusion kinetics in Carbon Onions was investigated for potential applications in batteries and electrochemical capacitors and we report a different loading behavior/voltage profile for these Carbon Onions as compared to graphitic systems. The Voltage vs Li concentration for Carbon Onions has a 'saw tooth' shape whereas the voltage profile for graphitic systems consists of plateaus.

(65f) Silicon-Rich Carbon Nanofibers From Water-Based Spinning to Control Combination of Silicon and Carbon for Li-Ion Battery

Joo, Y. L., Cornell University
Kim, Y. S., Cornell University
Kim, K. W., Cornell University
Cho, D.,
Hansen, N. S., Cornell University
Originally presented on: 11/4/2013 10:00:00 - 10:15:00

Silicon-rich carbon nanofibers from water-based spinning to control combination of silicon and carbon for Li-ion battery

Yong Seok Kim, Kyoung Woo Kim, Daehwan Cho, Nathaniel S. Hansen, and Yong Lak Joo*

School of Chemical and Biomolecular Engineering, Cornell University, Ithaca 14853, USA

ABSTRACT: Carbon (C) nanofibers with high silicon (Si) loading (> 65 wt%) were fabricated from water–based electrospinning of aqueous polyvinyl alcohol/Si nanoparticle solution for Li–ion battery anode applications. Our Si-rich C (SRC) nanofibers showed much facilitated charge transport features and increased activities because of continuous one–dimensional (1–D) carbon backbone structure, leading to superior battery performance to that of bare silicon nanoparticles. The presence of carbon as 1–D nanofibers could not only mitigate volume expansion of silicon but also avoid formation unstable solid electrolyte interface onto the surface of silicon. Our study on the optimum combination of carbon and silicon in the nanofibers for their electrochemical properties and battery performance revealed that the SRC nanofibers with 72.8 wt% Si content (27.2 wt% C) exhibits an adequate balance between high energy capacity of Si nanoparticles and dimensional stability with effective charge transport by carbon nanofibers, leading to an outstanding cycle life to maintain 1076 mAh/g normalized by total electrode mass and high Coulombic efficiency of about 99% for fifty cycles. Such scalable Si-rich C nanofibers made from water-based electrospinning using polyvinyl alcohol should be greatly contributed to a cost–effective development of high performance negative battery electrodes.

(65g) Electrochemical Properties of Carbon-Coated SnO2-Graphene Nanocomposites for Rechargeable Lithium-Ion Batteries

Ara, M., Wayne State University
Salley, D. S., Wayne State University
Ng, S. K. Y., Wayne State University
Originally presented on: 11/4/2013 10:15:00 - 10:30:00

SnO_2 nanoparticles were carbon-coated using two different methods and then dispersed on the reduced graphene oxide nanosheets prepared from modified Hummer's method. Carbon-coating was performed by a solvothermal method from glucose precursor material or by carbonization in Polyacrylonitrile (PAN) based carbon fibers.The later was prepared by dispersing SnO_2 nanoparticles in N-Methyl-2-pyrrolidone and then the dried powder was stabilized in air at 270°C for 30 minutes followed by the heat treatment at 900°C for 2 hours. The coated nanocomposites were characterized by X-ray diffraction, scanning electron microscopy (SEM), transmission electron microscopy (TEM), and Raman spectroscopy. From TEM observations, it was found that the SnO_2 nanoparticles were 2-6 nm in diameter and were uniformly distributed on the reduced graphene oxide nanosheets. Electrochemical properties were evaluated by galvanostatic charge-discharge cycling. Carbon-coated SnO_2-reduced graphene oxide nanocomposites showed higher capacity and more stable electrochemical performances than non-coated SnO_2-reduced graphene oxide nanocomposites. Carbon coating enhances the conductivity of the electrode and it reduces the stress developed by the volume changes of the SnO_2 nanoparticles during charge/discharge cycles. The synergistic effect of the amorphous carbon, SnO_2 nanoparticles and reduced graphene oxide nanosheets is also responsible for the superior specific capacity of carbon-coated anode material. Excellent rate capability and cyclic performances were also demonstrated with higher current densities of 500 and 1000 mA/g.

(65h) Multilayer Hybrid Electrodes of TiO2/Mwnt for Electrochemical Applications

Hyder, M. N., Massachusetts Institute of Technology
Kavian, R.,
Shao-Horn, Y., Massachusetts Institute of Technology
Hammond, P. T., Massachusetts Institute of Technology
Originally presented on: 11/4/2013 10:30:00 - 10:45:00

Rational design of next-generation electrochemical devices requires novel electrode materials and architectures to achieve high-energy, high-power density to bridge the gap between battery and capacitors. Conventional electrode fabrication processes such as doctor blading allows limited control over the electrode thickness and structure at the nanoscale, in addition to the incorporation of insulating binder materials that limits device performance. To address these challenges, we have engineered a route for the synthesis of highly stable, sub-8nm TiO_2 nanoparticles that were incorporated with acid-functionalized multi-walled carbon nanotubes (MWNT). Using electrostatic layer-by-layer nanofabrication, we assembled binder-free hybrid electrode of TiO_2/MWNT to obtain a synergistic effect from high electronic conductivity and excellent charge storage capacity. These thin film electrodes with highly controllable thicknesses shows well developed mesopores with finely dispersed, non-agglomerated TiO2 nanoparticles on MWNTs. Electrochemical measurements show high charge storage capacity (>150 mAh/$g_{electrode}$ at 0.1 A/$g_{electrode}$) with excellent cycling stability during charging and discharging up to 200 cycles indicating their promise as thin-film negative electrodes for future Li storage applications.https://aiche.confex.com/aiche/2013/t5/papers/index.cgi?username=341047&password=343273

(65i) Facile Preparation of Mesoporous α-Fe2O3 With Unique Cocoon-Like Hollow Nanostructure for Reversible Lithium Ion Storage

Zhu, J., Wayne State University
Ng, K. Y. S., Wayne State University
Deng, D., Wayne State University
Originally presented on: 11/4/2013 10:45:00 - 11:00:00

After 20 years of development since the first commercialization of rechargeable lithium ion batteries by Sony in 1991, carbon is still exclusively used as anode materials for lithium ion batteries. However, the theoretical capacity (372mAh g^{-1}) of graphite is almost achieved now and it becomes a bottleneck to further increase lithium-ion storage capacity using carbon. Carbon-alternative materials with higher capacity are currently being extensively explored to meet the increasing demand for energy in mobile electronic devices, electric vehicles and energy storage in grid. Transition metal oxides receive increasing attention as carbon-alternatives because of their much higher theoretical capacity. For example, α-Fe_2O_3 with a theoretical capacity of 1007 mAh g^{-1} is a promising candidate. Compared to other transition metal oxides, α-Fe_2O_3 is outstanding in terms of low cost, abundance, ease of fabrication, and environmental benignity. However, the poor cyclability caused by volume change during the insertion /extraction of Li ions and poor conductivity is the main challenge. Our strategy is to design and tailor α-Fe_2O_3 at nanoscale with structure control to improve their electrochemical performance. In this presentation, the strategy to facilely prepare mesoporous α-Fe2O3 with unique cocoon-like hollow nanostructure as anode materials will be reported. The formation mechanism is revealed by extensive characterization by FESEM, TEM and XRD for samples prepared under different controlled conditions. The improved electrochemical performance in reversible lithium ion storage will be discussed.

(134a) Identifying Defect Sites in Lithium-Ion Battery Materials: Local Disorder in LiVPO4F and Its Influence On Bulk Properties

Messinger, R. J., CNRS
Menetrier, M., CNRS
Carlier, D.,
Ateba Mba, J. M., CNRS
Croguennec, L.,
Masquelier, C.,
Massiot, D.,
Deschamps, M.,
Originally presented on: 11/4/2013 12:30:00 - 12:45:00

The transition metal fluorophosphate $LiVPO_4F$ is a promising electrode material for lithium-ion batteries that is capable of operating both as a high-voltage positive electrode (4.25 V, $V^{III/IV}$) and a negative electrode (1.8 V, $V^{II/III}$).[1,2] While its Tavorite-like structure has been solved by a combination of X-ray and neutron diffraction, solid-state nuclear magnetic resonance (NMR) measurements reveal a significant extent of local disorder that has not been observed at longer length scales with scattering techniques. In particular, solid-state single-pulse ^7Li NMR spectra acquired under conditions of ultra-fast (64 kHz) magic-angle-spinning (MAS) reveal that $LiVPO_4F$ exhibits several additional ^7Li NMR signals of unknown origin that can account for up to 20% of the total lithium content. Electrochemical measurements in Swagelok-type cells with lithium metal as the negative electrode establish that materials with different populations of unknown lithium sites exhibit different cycling behavior, including differing extents of irreversible capacity and polarization. 2D $^7Li\{^7Li\}$ dipolar correlation NMR measurements establish the sub-nanometer proximities of the unknown lithium sites with respect to the $LiVPO_4F$ site, proving unambiguously that the unknown sites are local defects within the $LiVPO_4F$ structure and are not associated with separate impurity phases. In combination with magnetic measurements, variable-temperature NMR measurements reveal that the paramagnetic ^7Li NMR shifts of the different lithium sites vary linearly with the bulk magnetic susceptibility, further confirming that the sites are defects within the $LiVPO_4F$ structure. The results provide a general strategy for identifying lithium defect sites in lithium-ion battery materials and highlight the importance of understanding local disorder and its relationship to macroscopic material and device properties.

[1] J.-M. Ateba Mba, C. Masquelier, E. Suard, L. Croguennec, *Chem. Mater.* **2012,** *24,* 1223 – 1234.

[2] J.-M. Ateba Mba, L. Croguennec, N.I. Basir, J. Barker, C. Masquelier, *J. Electrochem. Soc.* **2012,** *159,* A1171-A1175.

(134b) Electrochemical Properties and Characterization of Li1.2Mn0.54Co0.13Ni0.13O2 Composite Cathode Powders Prepared by Ultrasonic Spray Pyrolysis

Langrock, A., University of Maryland
Ehrman, S. H., University of Maryland
Wang, C., University of Maryland
Originally presented on: 11/4/2013 12:45:00 - 13:00:00

Lithium-rich layered oxide $Li_{1.2}Mn_{0.54}Co_{0.13}Ni_{0.13}O_2$ (LMO) composite cathode powders are prepared by ultrasonic spray pyrolysis. The crystal structure and morphology of the prepared samples are characterized with X-ray diffraction and scanning electron microscope. The spherical shape and porous nanostructure of the as prepared powders are maintained after post-treatment at temperatures between 600 and 800°C. Grain growth in the powders increases with increasing post-treatment temperatures. The BET surface area of the composite cathode powders post-treated at 800°C is 12 $m^2\ g^{-1}$. Discharge capacity of the cathode powders post-treated at 800°C decreases from 278 mAh g^{-1} to 208 mAh g^{-1} by the 50th cycle, the capacity retention being 75%. The post-treatment temperature strongly affected the electrochemical properties of the composite cathode powders. Composite cathode powders post-treated at high temperatures with high amounts of the Li_2MnO_3 phase had good cycle properties as well as large initial charge/discharge capacities.

(134c) Atomic Layer Deposition (ALD) On the Nanostructured Li-Mn-Rich Composite Li1.2Ni0.13Mn0.54Co0.13O2 Cathode Powder

Zhang, X., Institute of Nuclear and New Energy Technology, Tsinghua University
Belharouak, I.,
Originally presented on: 11/4/2013 13:00:00 - 13:15:00

Atomic Layer Deposition (ALD) on the Nanostructured Li-Mn-rich Composite $Li_{1.2}Ni_{0.13}Mn_{0.54}Co_{0.13}O_2$Cathode Powder

Xiaofeng Zhang,[a] Ilias Belharouak,[a,*]

[a] Chemical Sciences and Engineering Division, Argonne National Laboratory, 9700 S. Cass Ave, Argonne, IL 60439, USA, Email: belharouak@anl.gov(Corresponding author)

Li-Mn-rich (LMR) composite cathode materials $Li_{1+x}M_{1-x}O_2$ (M = Ni, Mn and Co) are of particular interest for plug-in hybrid electric vehicles (PHEVs), in which a high-energy battery is always desired for a higher mileage per charge.[1, 2] In the composite structure, the layered, the spinel and rocksalt structure are compatible at nano-scales, that many composites with different electrochemical properties can be synthesized by simply tuning the Li content in the materials. However, these materials would suffer from voltage decay and capacity fading with cycles, which significantly deteriorate their energy densities and life time for practical applications. In this type of materials, Li content is essential to the arrangement of the transition metal atoms in the composites and also to the structural stability against undesired reactions in a real electrochemical cell. It has been demonstrated that the occurrence of layered-spinel transformation induces the voltage decay. It is because the Li intercalation/deintercalation reactions may preferentially occurs at the tetrahedral site other than octahedral site, which facilitates the spinel formation.[1, 2] Further study by HR-TEM indicated that the composite particle surface could lose the layered feature after a few cycles, and the first-principle calculation suggested a defect-spinel structure forming on the surface due to the migration of transition metal ions into the Li layers.[3] Therefore, the surface chemistry and structure stability appear to be essential to limit the undesired reactions and phase transformation in the composite materials. Some achievements have been made *via* developing new synthesis processes, new chemistry or surface modifications (e.g., coating), which possibly altered the short-range and long-range ordering of transition metal atoms (Ni and Mn) near the surface, which lead to an improvement in the battery performance.[4-8]

Atomic layer deposition (ALD) is a well-established method to grow ultrathin films on flat surfaces and powders in atomic thickness using sequential, self-limiting surface reactions.[9-14] It has been demonstrated that the surface film can react as a surface barrier between the cathode active material and the electrolyte, which would reduce the loss of active materials in the electrode. [9-14] Consequently, less capacity loss is achieved for the coated cells as compared with the pristine materials. The well-established ALD process normally is using trimethylaluminium (TMA) and H_2O as precursors forming an Al_2O_3 surface film. Experimental results showed that the coated $LiCoO_2$ powders exhibited a capacity retention of 89% after 120 charge–discharge cycles in the 3.3–4.5 V vs. Li/Li+ range. [12, 15-17] The ALD surface coating can also be applied to battery electrodes at a low temperature <150 °C, which would not degrade the electrodes (e.g., melting of polymer binders) in this temperature range. It is reported that Al_2O_3-coated $LiCoO_2$ electrodes display superior electrochemical performance in comparison to the uncoated $LiCoO_2$ electrodes. However, Liu et al observed that the ALD surface coated Al_2O_3 could be lithiated forming passivation $LiAlO_2$ glass on the materials surface using in-situ TEM [18], which provided new insight into the electrochemical characteristic of the Al_2O_3surface film against Li.

In this paper, we are aiming for studying the importance of a conformal surface protection film using atomic layer deposition (ALD) technique on the high-capacity $Li_{1.2}Ni_{0.13}Mn_{0.54}Co_{0.13}O_2$ cathode materials in Li-ion batteries, which can be potentially significantly improved in their electrochemical performance. More precisely, we evaluated the effect of coating species, coating thickness and morphology on the rate performance, cycling life and high temperature performance of the high-capacity $Li_{1.2}Ni_{0.13}Mn_{0.54}Co_{0.13}O_2$ cathode materials in half-cell and full-cell configurations.[19] We found that an conformal Al_2O_3 thin film could be deposited on the powders using 10 ALD cycles of TMA and H_2O at 150 °C. TiO_2 coating was prepared using 20 ALD cycles of titanium isopropoxide (TTIP) and H_2O at 150 °C, which displayed a particulate morphology. Nevertheless, the surface coated TiO_2 film was electrochemically active, which could be lithiated forming Li_xTiO_2 as detected by X-ray photoelectron spectroscopy (XPS). The electrochemical reactivity of TiO_2 will cause degradation of the surface film, leading to capacity fading of the cell. Nevertheless, it was also found that the Al_2O_3 surface coating was not electrochemically active after repetitive cycles, which was contrary to the early report by Lui et al that Al_2O_3 can be lithiated into $LiAlO_2$. [18] To further explorate the phenomena, a systematic study on the ALD coating of Al_2O_3 and $LiAlO_2$was performed.

Refereances:

[1] C.S. Johnson, N. Li, J.T. Vaughey, S.A. Hackney, M.M. Thackeray, Electrochem. Commun., 7 (2005) 528-536.

[2] C.S. Johnson, N.C. Li, C. Lefief, J.T. Vaughey, M.M. Thackeray, Chem. Mater., 20 (2008) 6095-6106.

[3] B. Xu, C.R. Fell, M.F. Chi, Y.S. Meng, Energy & Environmental Science, 4 (2011) 2223-2233.

[4] J. Liu, B. Reeja-Jayan, A. Manthiram, Journal of Physical Chemistry C, 114 (2010) 9528-9533.

[5] S.H. Kang, M.M. Thackeray, J. Electrochem. Soc., 155 (2008) A269-A275.

[6] J.R. Croy, M. Balasubramaniam, D. Kim, S.-H. Kang, M.M. Thackeray, Chem. Mater., 23 (2012) 5415-5424.

[7] J.R. Croy, S.H. Kang, M. Balasubramanian, M.M. Thackeray, Electrochem. Commun., 13 (2011) 1063-1066.

[8] D. Kim, S.H. Kang, M. Balasubramanian, C.S. Johnson, Electrochem. Commun., 12 (2010) 1618-1621.

[9] M.Q. Snyder, S. Trebukhova, B. Ravdel, M.C. Wheeler, J. DiCarlo, C.P. Tripp, W.J. DeSisto, Mater Res Soc Symp P, 972 (2006) 331-338.

[10] M.Q. Snyder, S.A. Trebukhova, B. Ravdel, M.C. Wheeler, J. DiCarlo, C.P. Tripp, W.J. DeSisto, J. Power Sources, 165 (2007) 379-385.

[11] S.M. George, Chem. Rev., 110 (2010) 111-131.

[12] Y.S. Jung, A.S. Cavanagh, A.C. Dillon, M.D. Groner, S.M. George, S.H. Lee, J. Electrochem. Soc., 157 (2010) A75-A81.

[13] L.A. Riley, A.S. Cavanagh, S.M. George, S.H. Lee, A.C. Dillon, Electrochem. Solid-State Lett., 14 (2011) A29-A31.

[14] L.A. Riley, S. Van Ana, A.S. Cavanagh, Y.F. Yan, S.M. George, P. Liu, A.C. Dillon, S.H. Lee, J. Power Sources, 196 (2011) 3317-3324.

[15] Y.S. Jung, A.S. Cavanagh, L.A. Riley, S.H. Kang, A.C. Dillon, M.D. Groner, S.M. George, S.H. Lee, Advanced Materials, 22 (2010) 2172-2176.

[16] I.D. Scott, Y.S. Jung, A.S. Cavanagh, Y.F. An, A.C. Dillon, S.M. George, S.H. Lee, Nano Lett., 11 (2011) 414-418.

[17] Y.S. Jung, P. Lu, A.S. Cavanagh, C. Ban, G.H. Kim, S.H. Lee, S.M. George, S.J. Harris, A.C. Dillon, Advanced Energy Materials, 3 (2012) 213-219.

[18] Y. Liu, N.S. Hudak, D.L. Huber, S.J. Limmer, J.P. Sullivan, J.Y. Huang, Nano Lett., 11 (2011) 4188–4194.

[19] X. Zhang, I. Belharouak, L. Li, Y. Lei, J.W. Elam, A. Nie, X. Chen, R. Shahbazian-Yassar, R.L. Axelbaum, Advanced Energy Materials, (2013). 10.1002/aenm.201300269

(134d) Metal Fluoride/Graphene Composites for High-Performance Li-Ion Cathodes

Hayner, C. M., Northwestern University
Zhao, X., Northwestern University
Yu, Y. Y., Northwestern University
Kung, M. C., Northwestern University
Kung, H. H., Northwestern University
Originally presented on: 11/4/2013 13:15:00 - 13:30:00

Rechargeable batteries with improved energy and power density have many potential, high impact applications including advanced portable electronics and electric vehicles. However, most 'next-generation' materials capable of providing improved performance suffer from rapid capacity degradation or severe loss of capacity when rapidly discharged. Graphene, a unique two-dimensional material composed of atomically thin carbon sheets, has shown potential to address unsatisfactory rate capability, limited cycling performance and abrupt failure of these high capacity material. Therefore, we have been exploring graphene-based architectures, allowing us to exploit graphene's unique combination of high surface area, high in-plane electrical conductivity, excellent tensile modulus and mechanical durability.

Metal fluorides, including iron trifluoride (FeF_3), have the potential of both high energy delivery and high cyclability. We have recently developed a stable nanostructured iron fluoride (FeF_3)/graphene composite cathode. A facile route that combines co-assembly and photothermal reduction was developed to synthesize free-standing, flexible FeF_3/graphene papers. The papers contained a uniform dispersion of FeF_3 nanoparticles (< 40 nm) and open ion diffusion channels in the porous, conducting network of graphene sheets that resulted in a flexible paper cathode with high charge storage capacity, rate, and cycling performance, without the need for other carbon additives or binder. The composites demonstrate promising applications as cathodes in high-energy density Li-ion batteries. In a free-standing form, they showed a high storage capacity of >400 mAh/g and improved cycling performance compared to bare FeF_3 particles. In addition, we have investigated a novel mixed metal fluoride cathode system, containing tunable manganese and iron fluoride, for which characterization and electrochemical performance will be presented.

(134e) Nanostructured Li2S-Graphene Composites As Cathode for High Energy Density Li-S Batteries

Han, K., Northwestern University
Shen, J., Northwestern University
Hayner, C. M., Northwestern University
Yu, Y. Y., Northwestern University
Kung, M. C., Northwestern University
Kung, H. H., Northwestern University
Originally presented on: 11/4/2013 13:30:00 - 13:45:00

Lithium-ion batteries, although a promising technology for energy storage, is reaching their energy density limitation, making it difficult to fulfill the requirements of emerging applications such as electric vehicles and smart grid. Lithium-sulfur batteries are one of the most attractive systems for next generation high energy storage beyond Li-ion batteries. Conventional Li-S batteries mainly use elemental sulfur as the starting cathode material due to its high theoretical capacity of ~1675 mAh/g and high energy density of ~2500 Wh/kg based on the redox reaction $S_8 + 16Li \rightleftarrows 8Li_2S$. Sulfur is also low cost, abundant, and of low toxicity. However, significant hurdles such as rapid capacity fade and poor rate capability have to be overcome before practical application can be realized. Two important underlying reasons for the poor performance of this battery are the insulating nature of sulfur and the dissolution of the intermediate product polysulfide into the electrolyte resulting in redox shuttling. Although extensive research efforts have been expanded to solve these issues, most studies are limited to the simple cell configuration of sulfur as cathode and metallic lithium as anode. For safety and ease of operation, introducing Li into the cell in the form of Li_2S is more desirable than Li metal. Moreover, Li_2S, with a theoretical capacity of ~1166 mAh/g, could be paired with a lithium-metal-free anode such as Si or Sn.

Here, we report the use of a nanostructured Li_2S-graphene composite as cathode material. Li_2S nanoparticles (20~50nm) are imbedded in pocket-like structures in a three-dimensional graphene/reduced graphene oxide via a facile, wet chemical method followed by thermal treatment. Structural characterization by XRD and SEM provide clear evidence for the unique structure. The constructed 3-D graphene network could enable efficient use of Li_2S active material by providing excellent electrical contacts, offer long-term stability against loss of polysulfide and accommodate volume changes during charge and discharge processes. The as-prepared Li_2S-graphene nanocomposites exhibit high initial reversible capacity and good cycling performance. The effect of lithium polysulfide as additive of the electrolyte is also investigated, especially with respect to its contribution to the overall charge storage capacity. The structural and electrochemical characterization results of the nanocomposites will be presented.

(134f) Carbon/Sulfur Microspheres With Multi-Modal Pore Structures for Lithium-Sulfur Battery Cathodes

Zhao, C., University of Wisconsin-Milwaukee
Liu, L., University of Wisconsin-Milwaukee
Zhao, H., University of Wisconsin-Milwaukee
Li, Y., University of Wisconsin-Milwaukee
Originally presented on: 11/4/2013 13:45:00 - 14:00:00

Sulfur has received increasing attention as a cathode material for lithium-sulfur batteries due to its high theoretical specific capacity of 1675 mAh g-1, which is five times higher than that of conventional lithium-ion batteries. However, poor electronic conductivity of sulfur, dissolution of the lithium polysulfide intermediates and large volume expansion of sulfur during cycling limit the commercialization of Li-S batteries. In this work, we have demonstrated a cathode design with a novel structure made of multi-modal porous carbon microspheres. The tunable porous structure of sulfur-infiltrated carbon nanocomposites allows easy access to Li ion transport, inhibits loss of active sulfur material, and provides void space for sulfur expansion. An initial capacity of 1289 mAh g-1 at 1C and a Coulombic efficiency of 90% over 200 cycles are achieved.

(134g) Nanofiber-Based Cathode Electrocatalysts Supported On Carbon Substrates for Lithium-Air Rechargeable Batteries

Joo, Y. L., Cornell University
Yin, J., Cornell University
Kim, J.,
Originally presented on: 11/4/2013 14:00:00 - 14:15:00

Nanofiber-Based Cathode Electrocatalysts Supported on Carbon Substrates for Lithium-Air Rechargeable Batteries

*Jun Yin[a], Jangwoo Kim[a], and Yong Lak Joo[a]**

[a] School of Chemical and Biomolecular Engineering, Cornell University, Ithaca, New York 14853

In terms of usable energy density rivaling that of gasoline, rechargeable lithium-air batteries have attracted increasing interests presently. However, the practical capacity was highly affected by the effective catalyst materials in the air cathode in Li-air batteries. The demand for effective catalyst materials in the air cathode requires the design and synthesis multifunction catalysts for power density improvement. In this presentation, several types of metal oxide nanofibers (e.g., ZnO, Co_3O_4 etc) were synthesized by water-based electrospinning of highly loaded precursor solutions, followed by thermal treatment and their use as the cathode catalysts in single-cell measurements was described. In the meantime, the correlation between the various carbon-based substrates and the metal oxide fiber catalysts and how they synergistically influence the electrocatalytic activity will also be discussed. The metal oxide nanofibers are characterized using an array of techniques such as SEM, XRD, XPS, BET and Raman to determine their morphology and structural properties. Charging – discharging evaluation, cyclic voltammetric measurements in organic electrolytes were carried out to assess the mechanistic details as well.

(134h) Low-Temperature Dehydrogenation From Aminoborane Complexes Under Carbon Dioxide Atmospheres

Lee, J. W., The City College of New York
Xiong, R., CCNY
Zhang, J., The City College of New York
Originally presented on: 11/4/2013 14:15:00 - 14:30:00

Ammonia borane (AB) is a chemical hydride and one of attractive solid-state media for on-board hydrogen storage due to high hydrogen contents (19.6 wt. %) and low-temperature dehydrogenation compared to other metal hydrides. The hydrogen release from AB can be accelerated at the working temperature (~ 85 °C) of polymer electrolyte membrane fuel cell. However, this talk shows that CO_2-treatment can accelerate hydrogen release of AB polymer complex, polyaminoborane (PAB). Polyaminoborane (PAB) is the spent fuel of ammonia borane (AB, NH_3BH_3) thermolysis with around 1 mole hydrogen release per mol AB. Dehydrogenation of PAB by thermolysis generally commences at temperatures above 120 °C. However, we find gaseous CO_2 environments dramatically enhance the thermolytic dehydrogenation of polyaminoborane (PAB) at 85 - 90 °C. By exposing PAB to 2.07 MPa CO_2, 5.48 wt % hydrogen based on the mass of ammonia borane (AB) desorbs in less than 1 h, while 4 h is required to generate 5.04 wt % hydrogen by maintaining PAB at 120 °C without using CO_2. The total hydrogen desorbed from the initial AB is 11.38 wt % under the CO_2 atmosphere at 85 °C, including the hydrogen release during the transition from AB to PAB. The main contributing factors to the facile hydrogen release are the promoting effect of newly formed formic acid and the exothermic effect of the reaction between PAB and CO_2. We will explain the possible promoting mechanism by various spectroscopic and elemental analyses.

(134i) Intermittent Electrical Energy Storage By Liquid Organic Hydrogen Carriers

Arlt, W., University of Erlangen-Nuremberg
Muller, K., University of Erlangen-Nuremberg
Teichmann, D., University of Erlangen-Nuremberg
Originally presented on: 11/4/2013 14:30:00 - 14:45:00

The intermittent delivery of energy from sources like photovoltaics or wind demands a large and long-term storage of energy. Liquid Organic Hydrogen Carriers like "Carbazol" address this topic using the infrastructure of liquids like domestic fuel oil. In Germany, the situation is far more complicated, because the country decided to phase out nuclear energy in the year 2022. The lecture highlights the situation in the German state of Bavaria and the planning.

The lecture reports the progress of Liquid Organic Hydrogen Carriers (LOHC). A demo-project is under construction at a condo in the city of Erlangen/Bavaria. Important data are safety, efficiency of electrical energy storage and robust construction. The purpose of this demo is to investigate

1. The ratio between storage and heat demand of the building
2. The number of storage and delivery actions
3. The robustness of the process in a public area

The time frame is 2 years from beginning of 2014 on.

(134j) Synthesis of Nanoporous Palladium Powder With Controlled Pore and Particle Size for Hydrogen Storage Applications

Cappillino, P., Sandia National Laboratories
Jones, C.,
Hattar, K., Sandia National Laboratories
Clark, B.,
Hekmaty, M.,
Jacobs, B.,
Robinson, D. B., Sandia National Laboratories
Originally presented on: 11/4/2013 14:45:00 - 15:00:00

Synthesis of Nanoporous Palladium Powder with Controlled Pore and Particle Size for Hydrogen Storage Applications

by Patrick J. Cappillino, Christopher G. Jones, Khalid M. Hattar, Blythe G. Clark, Michelle A. Hekmaty, Benjamin W. Jacobs, and David B. Robinson

Materials capable of rapidly storing and delivering hydrogen are currently in demand for numerous energy applications.Â The high surface area exhibited by nanoporous palladium has the potential to greatly improve its kinetics of hydrogen charge and discharge *versus* nonporous metal while retaining favorable operating temperatures, pressures, and volumetric capacity.Â In addition, the void space imparted by porosity should accommodate the volumetric expansion of palladium upon hydriding, mitigating plastic deformation and improving cycle life.

By chemically reducing palladium salts in the presence of various surfactants, we are able to synthesize palladium having size-tunable mesopores from a few to tens of nm, in a scalable fashion.Â Our results suggest that Pd nanoparticles form and sinter around micelles present in the aqueous media.Â Changing the chemical composition of the surfactant affects the size of these micelles and ultimately determines the pore dimensions.Â We demonstrate that pore size affects not only surface area, but also pore thermal stability under vacuum, and in the presence of hydrogen gas.

In addition to pore geometry, particle size and shape are important factors in determining gas flow characteristics. Using nonporous copper particles as sacrificial reductants for palladium salts, in the presence of surfactants, we are able to synthesize micron-sized particles of uniform size and shape.Â This is in contrast to continuous aggregates and films that are obtained using soluble chemical reductants or planar electrodes.

We will present details of the synthesis of nanoporous palladium, demonstrating control of both pore and particle size.Â Bulk measurements of hydrogen storage properties and pore characteristics, as well as microscopic measurements, including *in situ* heated-stage TEM under vacuum and in the presence of hydrogen, will be included, as well as details on the kinetics of hydriding/dehydriding.

Sandia National Laboratories is a multi-program laboratory managed and operated by Sandia Corporation, a wholly owned subsidiary of Lockheed Martin Corporation, for the U.S. Department of Energy's National Nuclear Security Administration under contract DE-AC04-94AL85000.

(192a) Fabrication of Porous Carbon Nanofibers With Adjustable Pore Sizes As Electrodes for Supercapacitors

Tran, C., Drexel University
Kalra, V., Drexel University
Originally presented on: 11/4/2013 15:15:00 - 15:45:00

We report a facile method for obtaining extremely high surface area and uniformly porous carbon nanofibers for supercapacitors. Blends of polyacrylonitrile and a sacrificial polymer at different compositions have been electrospun into non-woven nanofiber mats with diameters in the range of 200–400 nm. Electrospun nanofiber mats are then subjected to carbonization to obtain porous carbon nanofibers (CNFs) as polyacrylonitrile converts to carbon and the sacrificial polymer decomposes out creating intra-fiber pores. Resultant porous CNFs exhibit specific surface area of up to 1600 $m^2 g^{-1}$ with a large fraction of mesopores (2–4 nm). No additional chemical or physical activation process was used. We demonstrate the tunability of the pore sizes within CNFs by varying the composition. The non-woven fiber mats of porous CNFs are studied as free-standing electrode materials for supercapacitors eliminating the need for polymeric binding agents. Electrochemical measurements showed large specific gravimetric and volumetric capacitances of up to 210 F g^{-1} and 60 F cm^{-3} in 1 M H_2SO_4 at a high cyclic voltammetry scan rate of 100 mV s^{-1} due to the large fraction of mesopores. These materials retain 75% performance at a large current density of 20 A g^{-1} indicating excellent power handling capability.

(192b) Enhanced Energy Storage By Tunable Electrolyte Confinement in Structure-Directed CNT Arrays

Nicol, K. T., Mainstream Engineering Corporation
Zastrow, D., Mainstream Engineering Corporation
Hill, J. J., Mainstream Engineering Corporation
Originally presented on: 11/4/2013 15:45:00 - 16:00:00

The demand for power and energy dense storage devices is growing rapidly. While electrochemical double layer capacitors (EDLCs) offer high power density, many believe that high energy density is also obtainable through precise tuning of the electrolyte, electrode and structure of the overall cell. Multiwalled carbon nanotubes (MWNTs) are a promising electrode material for EDLCs because of their high electrical conductivity, chemical stability and exceptionally high surface area. Until single walled carbon nanotubes can be produced as solely conductive structures, dense and low-diameter MWNTs offer the best properties with regard to carbon-based EDLC electrode materials. When fabricated in anodic aluminum oxide templates, a MWNT is grown within each nanopore, resulting in an ultra-high density aligned array. This method produces a carbon electrode material with a significantly increased active surface area over commercially available materials. The energy density and capacitance of these materials can be further improved by tuning the size of the electrolyte domain around each MWNT. This is accomplished through partial and selective removal of the template, thereby confining the electrolyte domain. In typical EDLCs, the diffuse part of the double layer does not contribute to charge storage. However, we have observed an apparent charge storage enhancement in domain-confined MWNT electrodes. We will also discuss the tunable approach with regard to optimizing energy storage independent of the choice of REDOX couple or solvent.

(192c) Multifunctional Nitrogen-Rich â€œBrick-and-Mortarâ€ Carbon As High Performance Supercapacitor Electrodes and Oxygen Reduction Electrocatalysts

Yu, D.,
Wei, L., Nanyang Technological University
Chen, Y., Nanyang Technological University
Originally presented on: 11/4/2013 16:00:00 - 16:15:00

Novel porous carbon materials with excellent electronic, chemical and structural properties and high nitrogen content are desirable for many applications. Here, we show the design and synthesis of a new multifunctional porous carbon material with unique "point to plane" architecture through a simple but effective activation-free procedure. Porous carbon derived from a new type of copolymer poly (vinylidene chloride-co-acrylonitrile) serves as the nitrogen-rich "mortar". Reduced graphene oxide layers work as "bricks" with an aim to provide an open nanoscale scaffold and connect porous carbon, as well as modulate the ratio between mesopores and micropores. This new material has large surface area (957 m^2/g), high nitrogen content (6.6 at %), excellent conductivity (up to 5.1 S/cm), and favorable hierarchical meso- and microporosity. Benefiting from these intriguing features, this material shows ultrahigh specific capacitance at 361 F/g in aqueous electrolyte, far superior to that of many carbonaceous materials reported so far. The as-assembled asymmetric supercapacitors with the designed carbon material as negative electrodes and porous cobalt oxide nanorods as positive electrodes deliver a high-energy of 50.1 Wh/kg with a cell voltage of up to 1.6 V. Further, this material shows a superb electrocatalytic activity for oxygen reduction reaction in alkaline medium comparable to that of 20 wt% platinum-carbon electrodes, with much better methanol tolerant and long-term durability. We expect that this uniquely designed nitrogen-rich porous carbon material with its simple and scalable synthesis method will have great potentials for various applications in energy storage, energy conversion and catalysis.

(192d) Multiwalled Carbon Nanotubes With Tuned Surface Functionalities for Electrochemical Energy Storage

Wei, H., Lamar University
Gu, H., Lamar University
Guo, J., Lamar University
Wei, S., Lamar University
Guo, Z., Lamar University
Originally presented on: 11/4/2013 16:15:00 - 16:30:00

In the paper, surface functionalities effect on the electrochemical properties and energy storage properties of multiwalled nanotubes (MWNTs) were investigated via two different functionalized MWNTs, one modified with ether and carboxylic acid functional groups (denoted as MWNTs-EC), and the other with ester functional groups (denoted as MWNTs-E) as the supercapacitor materials. The functionalities onto the nanotubes were confirmed by Raman spectra and caused a minor enhancement in the band gap obtained from the Uv-vis spectra. Cyclic voltammetry results reveal that the capacitance of the MWNTs-EC, and MWNTs-E, was significantly increased by 41.5% and 123.2%, respectively, when compared to the as received MWNTs at lower scan rates. The energy density is also found to increase by 36.1 and 130.0% for the former and the latter, respectively, than that of the as received MWNTs-As in the galvanostatic charge/discharge measurement. The increases in both the capacitance and the energy density might arise from the enhanced surface wettability and larger pseudocapacitance contributions as confirmed by the electrochemical impedance spectroscopy. An equivalent circuit was proposed to disclose the nature of the capacitive behavior of the electrodes. The cyclic stability studies show an excellent long term cycling stability of the MWNTs, which retain 100.0% of the charge storage or discharge capacity even after 1000 charge-discharge cycles.

(192e) Three-Dimensional Core@Shell Nanostructured Array for Microscale Electrochemical Energy Storage

Mao, Y., Washington State University
Sun, X., Kansas State University
Originally presented on: 11/4/2013 16:30:00 - 16:45:00

Microelectromechanical systems (MEMS) based autonomous and biomedical devices play an ever increasing role in our society by monitoring the structural health of major civil works, the environmental presence of toxic or explosive compounds, and the physical health of human patients. Integration of robust and high power energy storage devices within these systems is important for reliable MEMS with necessary burst energy for data relay and dense components by reducing bulky external power sources. To overcome the shortcomings of conventional two-dimensional battery and supercapacitor configurations while developing more powerful energy storage devices, we focus on the development of three-dimensional architectures to take advantage of the vertical dimension-height. In this study, we designed a novel architecture composed of branched nanowire arrays, i.e. forest of nanotrees, to further boost electrochemical performance over one-dimensional nanowire arrays, inspired by branched trees with larger surface area to capture more sunlight. The prepared three-dimensional $ZnO@MnO_2$ core@shell branched nanowire arrays exhibit five times higher areal capacitance, better rate performance and smaller inner resistance than their nanowire array counterparts. These novel three-dimensional architectures offer promising designs for powering microelectronics and other autonomous devices on exceptionally small geometric scales.

(192f) Microfabricated Nickel Oxide Supercapacitors Based On High Aspect Ratio Concentric Cylindrical Electrodes

Armutlulu, A., Georgia Institute of Technology
Bidstrup Allen, S. A., Georgia Institute of Technology
Allen, M. G., Georgia Institute of Technology
Originally presented on: 11/4/2013 16:45:00 - 17:00:00

Supercapacitors are known as high-power energy storage devices, which narrow the application gap between the batteries and traditional electrostatic capacitors. As in the case of their electrostatic counterparts, they possess rapid charge and discharge capabilities with high degree of reversibility. Yet they can also store more energy per unit mass and/or volume which makes them superior to traditional capacitors and thus, yields a more attractive candidate for energy storage systems to be utilized in portable electronic devices. Supercapacitors can store energy electrostatically via separated charges in a Helmholtz double-layer that forms at the electrode/electrolyte interface, as well as electrochemically based on a phenomenon called pseudocapacitance that relies on reversible Faradaic charge transfer between the electrolyte and the surface of the electrode. In both energy storage mechanisms, electrode surface area plays a crucial role. In this study, our focus is on MEMS-enabled nickel oxide (NiO) based supercapacitors comprising high surface area 3D microelectrodes. The fabrication process consists of four major steps: formation of the array of high aspect ratio SU-8 pillars via backside exposure, deposition of alternating Cu and Ni layers on the SU-8 pillars by a sequential robotic electroplating process, mechanical and chemical etching of sacrificial Cu layers to obtain high surface area Ni structures in the form of concentric cylinders, and lastly conformal electrodeposition of the active material, i.e. nickel hydroxide ($Ni(OH)_2$), on the high surface area Ni backbone. Following the electrodeposition process of the $Ni(OH)_2$ layer, the structure is heated up to 300 °C in order to convert $Ni(OH)_2$ film to NiO, and the fabrication process is finalized. Performance tests for the resulting structures are carried out via cyclic voltammetry in 1 M KOH solution. Devices are cycled in excess of 500 times, and no significant capacitance degradation is observed. The areal and specific capacitance values of the supercapacitors are estimated to be 270 mF/cm^2 and 56 F/g, respectively.

(192g) Pretreatment Effects On Charge Storage of Early Transition-Metal Carbides and Nitrides

Djire, A., University of Michigan
Pande, P., University of Michigan
Sleightholme, A. E. S., University of Michigan
Deb, A., University of Michigan
Rasmussen, P. G., University of Michigan
Penner-Hahn, J., University of Michigan
Thompson, L. T., University of Michigan
Originally presented on: 11/4/2013 17:00:00 - 17:15:00

Pretreatment Effects on Charge Storage of Early Transition-Metal Carbides and Nitrides

Abdoulaye Djire[a], Priyanka Pande[a] , Aniruddha Deb[b], Alice E. S. Sleightholme[a], Paul Rasmussen[ac], James Penner-Hahn[b], Levi T Thompson[ac]*

[a] Department of Chemical Engineering

[b] Department of Chemistry

[c] Hydrogen Energy Technology Laboratory

University of Michigan, Ann Arbor, MI_- 48109-2130

E-mail: ltt@umich.edu

Early transition-metal carbides and nitrides are promising candidates for use in supercapacitor electrodes due to their high electronic conductivities, high surface areas (can exceed 200 m^2/g), good electrochemical stabilities and high capacitance [1,2]. For example, the capacitance for VN has been reported to be as high as 1340 Fg^{-1} in aqueous KOH [3]. This high capacitance has been attributed to pseudocapacitive charge storage involving fast near-surface redox reactions. We observed that pretreatment of the surface to remove the passivation layer significantly improved the capacitances and electrochemical stabilities of early transition-metal carbides and nitrides in aqueous electrolytes. In this paper we present results from electrochemical characterization including cyclic voltammetry (CV), chronopotentiometry, and electrochemical impedance spectroscopy (EIS), and surface characterization using x-ray photoelectron spectroscopy (XPS) to explain changes in surface chemistry and electrochemical properties caused by pretreatment.Â

The early transition-metal carbides and nitrides of V, W and Mo were synthesized via temperature-programmed-reaction of their oxide precursors with 15% CH_4/H_2 or anhydrous NH_3 followed by passivation in 1% O_2/He at room temperature to form a oxygen-rich passivation layer preventing bulk oxidation on exposure to air [1]. Characterization of the structural properties was performed using nitrogen physisorption (BET surface area) and X-ray diffraction. Prior to the electrochemical characterization, the electrodes were pretreated in 0.3M NaOH aqueous solution for 1 min then rinsed with acetone, ethanol, and ultrapure water [4]. The CV was used to establish the stability windows and capacitances for these materials. The capacitance was deconvoluted into double-layer capacitance and pseudocapacitance by means of CV and EIS. Chronopotentiometry was used for further stability measurements and to determine changes in surface activities caused by pretreatment.

As shown in Figure 1, the specific capacitance for VN increased by 57% after pretreatment. This enhancement in performance suggests significant improvement in material electrochemical activities. We believe that pretreatment altered the charge storage for the carbides and nitrides. We also observed good electrochemical stability after successive cycling of the material for several thousand cycles. In general, the treated materials were more stable than the passivated materials.

Figure 2 shows the open circuit potential (OCP) or rest potentials for the passivated and treated VN in acidic medium. The OCP for the treated material is lower than that for the passivated material. We attributed this to changes in surface activities due to pretreatment.Â Using the Nernst equation, we found that the treated VN had a higher pseudocapacitance than the passivated VN. This pseudocapacitive charge-storage mechanism is being investigated using XPS and *in-situ* x-ray absorption spectroscopy.

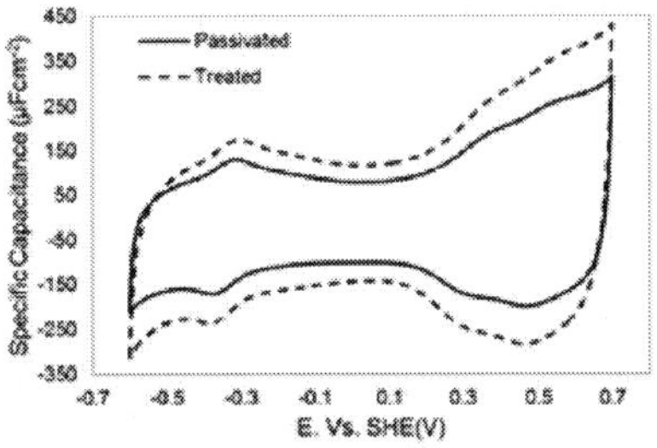

Figure 1: CV of passivated and treated VN in 0.1M H_2SO_4 at 50 mVs^{-1}.

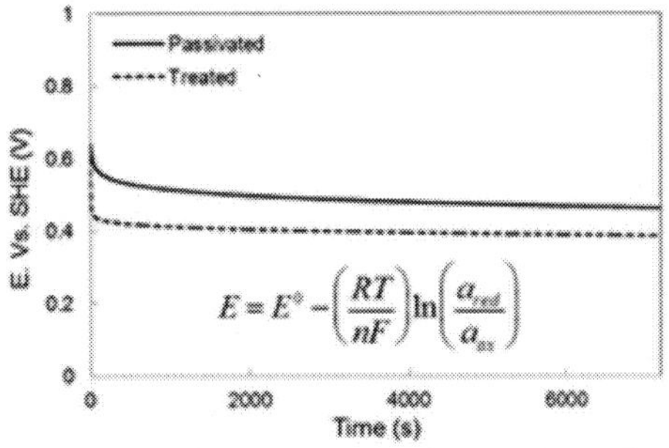

Figure 2: Rest potential of passivated and treated VN in 0.1M H_2SO_4

The physical and electrochemical properties will be described in terms of a proposed mechanism for psuedocapacitive charge storage.

References

(1) Cladridge, J. B.; York, A. P. E.; Brungs, A. J.; Green Malcolm L. H.; *Chem. Mater.* **2000**, 12, 132.

(2) Wixom, M.R.; Tarnowski, D. J.; Parker, J. M.; Lee , J.Q.; Chen, P. ?L.; Song, I.; Thompson, L. T.; *Mat. Res. Soc. Symp. Proc.* **1998,** 496, 643.

(3) Choi, D.;Â Kumta, P. N.; *Electrochem. Solid-State Lett.* **2005,** 8, 8, A418.

(4) Weidman, M.C.; Elposito, D.V.; Hsu, I. J.; Chen J. G.; J. Electrochem Soc.**2010**, 157, 12, F179

(192h) Doped Transition Metal Oxide Composite Electrodes for Supercapacitor Applications

Jampani Hanumantha, P., University of Pittsburgh
Kadakia, K., University of Pittsburgh
Hong, D. H., University of Pittsburgh
Poston, J., National Energy Technology Laboratory, U.S. Department of Energy
Ayyakkannu, M., National Energy Technology Laboratory
Kumta, P. N., University of Pittsburgh
Originally presented on: 11/4/2013 17:15:00 - 17:30:00

There is a dire need to provide a sustainable solution to the growing global demand for energy. It is this incessant demand that is the driving force fueling various current and emerging technologies including lithium ion batteries, supercapacitors and fuel cells. Supercapacitors occupy a unique place among these energy storage devices on account of their high power density and long cyclability. Supercapacitors are electrochemical charge storage devices capable of delivering large bursts of charge over a very short period of time. It is due to this unique ability that supercapacitors assume a very important role in the growing area of alternative energy storage systems. High surface area carbons, hydrated ruthenium oxide and MnO_2 are among the primary materials being considered for supercapacitor applications. Bulk oxide materials are traditionally hamstrung on account of their limited electrical conductivity for various electronic and electrochemical applications. In order to gain a fundamental understanding of the charge storage mechanisms in oxide materials, we deposited films of undoped transition metal oxides and doped transition metal oxides on current collectors and used various materials and electrochemical characterization techniques including X-ray photo-electron spectroscopy (XPS), cyclic voltammetry and electrochemical impedance spectroscopy to obtain a first-hand understanding of the underlying mechanisms. Effect of coating the various oxides on vertically aligned carbon nanotubes was also examined. Results of these studies will be presented and discussed.

(192i) Novel 3-D MnO2/Holey Graphene Nanostructure for Supercapacitor Applications With Enhanced Electrochemical Performances

Wang, L., Wayne State University
Meng, T., Wayne State University
Ara, M., Wayne State University
Deng, D., Wayne State University
Ng, S., Wayne State University
Originally presented on: 11/4/2013 17:30:00 - 17:45:00

As an excellent energy storage device with high-power density and long life cycle, supercapacitor has been intensely investigated in many aspects, such as electrode materials, electrode structure engineering, electrolyte compositions, separator improvements, as well as device design. Among them, one of the most urgent shortcuts is the around 10 times lower energy density compared to the traditional Li-ion batteries. This drawback has dramatically confined the applications of supercapacitors in electrical vehicles, intermittent energy storages and other occasions where both the power density and energy density are required. However, many researchers have been working on improving the energy density without significantly scarifying power density. These researches include increasing the voltage windows by employing organic electrolyte with larger voltage openings, applications of thinner and yet still mechanically strong separators, integrate materials with higher intrinsic capacitance, and so on. Despite all the effort on supercapacitor research, the current state-of-art on energy density still needs to be improved to meet the fast-growing application demanding in the emerging and potential applications. Therefore, we propose a mild method of preparing a novel 3-D MnO2/holey graphene composite for supercapacitor electrodes. The nano MnO2 nanorods in the composites serves as a capacitance booster due to its high intrinct capacitance and spacer to prevent 2-D graphene sheets from stacking back together. At the same time, graphene also serves both as active material to provide large surface area for charge storage and conductive agent to collect electrons generated by the surface redox reactions of semiconductive MnO2. Holes on graphene sheets enhance the efficient accessibility of electrolyte to the surface of active materials. With this novel 3-D composite, an increasement of around 30% to 50% in capacitance is expected.

(192j) Electrochemical Energy Storage of Magnetic Carbon Nanocomposites: Role of Magnetocapacitance and Magnetohydrodynamics

Zhu, J., Lamar University
Wei, S., Lamar University
Guo, Z., Lamar University
Originally presented on: 11/4/2013 17:45:00 - 18:00:00

Magnetic field has been used in different chemical processes to improve the quality of electrodeposited films, yield of photoredox products, and oxidative reactivity of catalysts and *etc*. In this work, a small external magnetic field will be imposed on a supercapacitor unit trying to improve its electrochemical energy density. Magnetic microtubular carbon nanocomposite fabrics decorated with uniformly distributed metal oxides (iron, cobalt and nickel oxides) nanoparticles will be fabricated via both microwave-assisted and conventional annealing processes. These fabrics could serve as flexible electrodes for supercapacitors to deliver high energy density and exhibit excellent cycling stability. The effect of metal oxide species, annealing method and applied magnetic field on the capacitance of each nanocomposite material will be tested using cyclic voltammetry and galvanostatic charge/discharge measurements. The corresponding mechanism will be investigated using electrochemical impedance spectroscopy. Cycling stability of each composite with and without magnetic field will be also comparatively studied.

(192k) Synthesis and Characterization of Orange Peel As Electrodes in Li-Ion Capacitors

Arie, A. A., Parahyangan University

Originally presented on: 11/4/2013 18:00:00 - 18:15:00

Orange peel was utilized as the raw material for activated carbon-based electrodes in Li-ion capacitors. The carbon samples weres synthesized
using a chemical activation process followed by a surface modification method. The surface modification method showed no significant effect on the specific surface area, however it changed the surface property of the carbon samples. Experimental results demonstrated that the electrochemical characteristics of un-treated activated carbons had a better specific capacitance compared with those obtained by surface modification method. Based on our results, orange peel waste can potentially be used as a cheap raw material for the mass production of activated carbons as electrodes in Li-ion capacitors.

(262a) Effects of Composition and Compositional Distribution On the Optoelectronic Properties and Function of Semiconductor Ternary Quantum Dots

Han, X., University of Massachusetts Amherst
Pandey, S. C., University of Massachusetts
Maroudas, D., University of Massachusetts Amherst
Originally presented on: 11/5/2013 8:30:00 - 8:49:00

The function of nanometer-size quantum dots (QDs) of ternary compound semiconductors, such as $In_xGa_{1-x}As$, $ZnSe_{1-x}S_x$, and $ZnSe_{1-x}Te_x$, used in the fabrication of optoelectronic and photovoltaic devices can be optimized by precise tuning of their electronic band gap through control of the QD composition (x) and diameter. In this presentation, we report results on compositional distributions in ternary QDs and how they affect the QDs' electronic band gap. We follow a hierarchical modeling approach that combines first-principles density functional theory (DFT) calculations and classical Monte Carlo simulations with a continuum model of species transport in spherical nanocrystals. In certain cases, the model predicts the formation of a concentration boundary layer near the nanocrystal surface, i.e., the formation of core/shell-like structures with shell regions rich in the surface segregating species. A systematic parametric analysis generates a database of transport properties, which can be used to design post-growth thermal annealing processes that enable the development of thermodynamically stable QDs.

Based on the continuum transport model, we have analyzed species interdiffusion kinetics in ternary semiconductor nanocrystals during their thermal annealing and used the modeling results to interpret the evolution of near-surface species concentration according to X-ray photoelectron spectroscopy (XPS) measurements. We have also explored the impact of composition and compositional distribution on the ternary QDs' electronic band gaps based on first-principles DFT calculations and found that ternary QDs with thermodynamically stable compositional distributions allow for precise band-gap tuning. Our findings lead to a proposal for an efficient one-step ternary QD synthesis method followed by annealing to promote self assembly of the thermodynamically stable configuration for optimal optoelectronic function in devices.

(262b) Molecular Modeling of Nanoparticles and Conjugated Polymers During Synthesis of Photoactive Layers of Organic Photovoltaic Solar Cells

Mortuza, S.,
Cisneros, C.,
Dela Cruz Bartolo, M.,
Banerjee, S.,
Originally presented on: 11/5/2013 8:49:00 - 9:08:00

Molecular modeling of nanoparticles and conjugated polymers during synthesis of photoactive layers of organic photovoltaic solar cells

S.M. Mortuza (1), Corinna Cisneros (1), Mark Dela Cruz Bartolo (2) and Soumik Banerjee (1)

1. School of Mechanical and Materials Engineering, Washington State University, Pullman, Washington, 99164, USA

2. California State University San Marcos, San Marcos, CA 92096

Solar cells are widely considered as potential alternative energy devices that can meet the increasing global energy demand. In particular, organic photovoltaic (OPV) solar cells, which employ organic nanoparticles in conjugated polymers as photoactive layers for energy conversion, have generated significant scientific interest due to their ease of manufacturing, relative low costs and flexibility. However, the power conversion efficiency (PCE) of OPV solar cells is still limited to 10-12% in part due to recombination of excitons before they can reach a nanoparticle-polymer interface. Such losses can be minimized by synthesizing photoactive layers with specific morphologies that result in large contact surface area between polymer and nanoparticles in the photoactive layer achieved through solution-based synthesis process. The scientific literature lacks detailed study of the effect of synthesis parameters, such as choice of solvent, processing temperature and relative concentration of nanoparticles and polymers, on the morphology of the photoactive layer. In order to understand the effects of these synthesis parameters on the agglomeration of nanoparticle, we performed molecular dynamics (MD) simulations of binary systems that comprise [6,6]-phenyl-C61-butyric acid methyl ester (PCBM), a commonly used acceptor, and various solvents in our previous study. The study suggested that indane is a better choice of solvent than other commonly used solvents for fabricating the photoactive layer of OPV solar cells. A systematic study of the effect of temperature and concentration of PCBM on agglomerate formation of PCBM showed that the extent of agglomerate formation increases with increasing temperature and concentration.
In an effort to further determine the agglomeration of PCBM in the conjugated polymer, we simulated systems comprising PCBM in poly-hexylthiophene (P3HT) that mimic the structure of a realistic

photoactive layer after spin coating. We evaluated the agglomeration behavior of PCBM by determining sizes of the largest clusters of PCBM in P3HT. To obtain further insight into the agglomerate structure of PCBMs, we evaluated the radial distribution functions (RDFs), which describe the coordination of one group of atoms or molecules, such as PCBMs or P3HTs, with the same or other group of atoms/molecules in the agglomerate. Our results indicate that the solubility of PCBM is greater in P3HT than in indane. Our simulations also demonstrate that PCBMs have a tendency to agglomerate in P3HT and the extent of agglomerate formation increases with increasing concentration of PCBM. We also analyzed the orientation of P3HTs to determine the extent of crystallinity. The results show that P3HTs have greater likelihood of parallel stacking in the system comprising PCBM and P3HT in weight ratio 1:2 than in the system that comprise 1:1 weight ratio of PCBM and P3HT. Figures 1 (a) and (b) show a magnified snapshot of a subsection of the simulation domain that illustrates the molecular arrangement of PCBM and P3HT in systems comprising PCBM and P3HT in weight ratios 1:1 and 1:2 respectively at 310 K. The aggregation pattern of PCBM and P3HT in Figure 1(a) demonstrates a greater contact area of PCBM and polymer compared to that in Figure 1 (b). Overall, our results show that relatively low concentration of PCBM in P3HT leads to sparse networks for charge transfer. In contrast, greater concentration of PCBM in P3HT leads to formation of suitable networks for transporting holes to photoanode and electrons to photocathode respectively. Results presented in this study provide fundamental insight that can help in selecting favorable processing parameters during synthesis of photoactive layer for OPV applications.

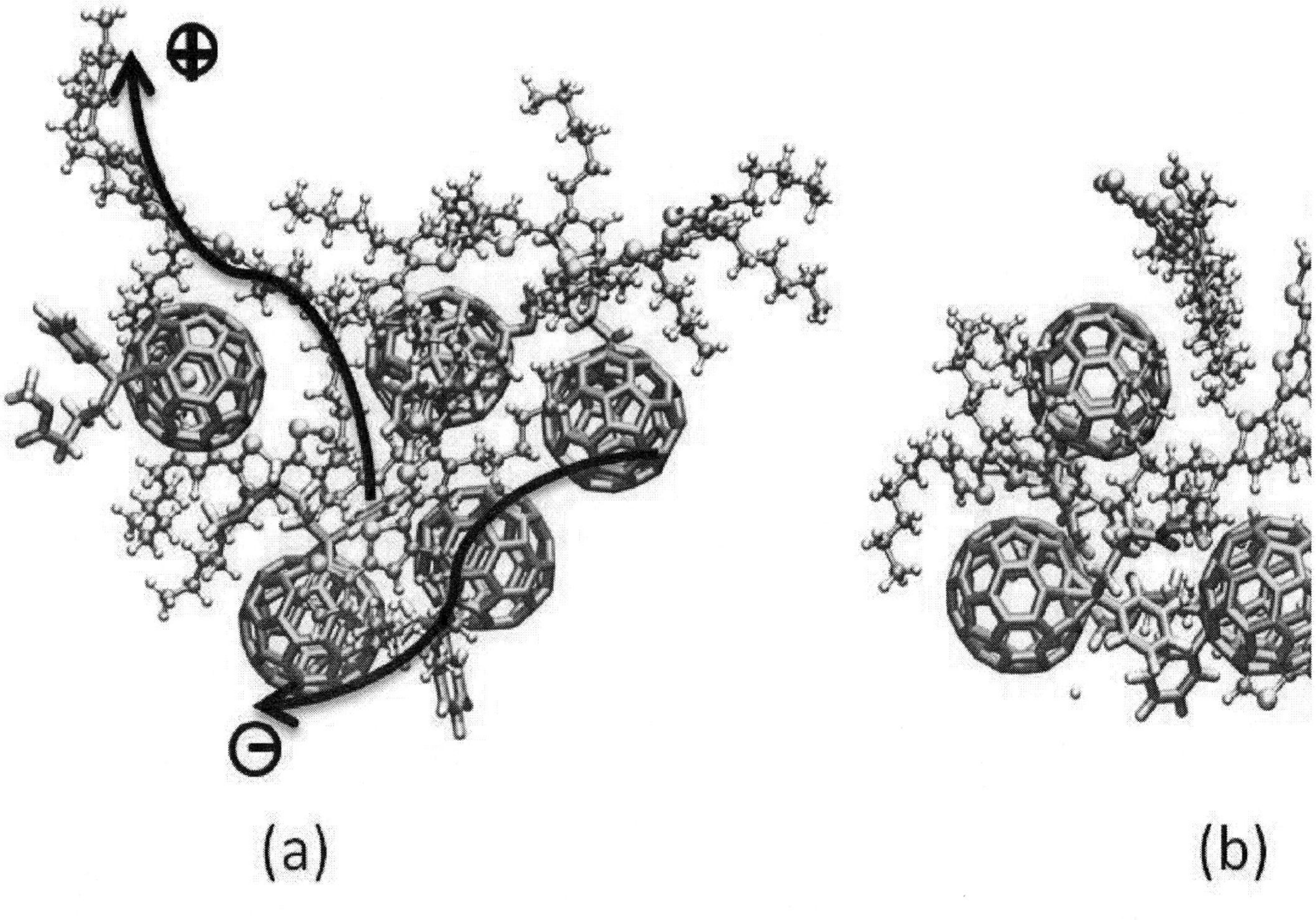

Figure 1: The molecular arrangement of PCBM and P3HT in systems of (a) 1:1 PCBM-P3HT and (b) 1:2 PCBM-P3HT at 300 K.

(262c) Integrating Photosystem I Proteins With Advanced Materials for Biologically Inspired Solar Energy Conversion

Jennings, G. K., Vanderbilt University
LeBlanc, G., Vanderbilt University
Gunther, D.,
Yang, S., Vanderbilt University
Cliffel, D., Vanderbilt University
Originally presented on: 11/5/2013 9:08:00 - 9:27:00

Photosystem I (PSI) is a ~10 nm protein complex that drives photosynthesis. Because of its nanoscale size and photodiode-like properties, PSI has drawn a surging interest for use in biologically inspired energy conversion devices upon extracting it from plants and assembling it as films on electrode surfaces. The research described in this presentation will focus on the molecular and electronic integration of spinach-extracted PSI at electrode surfaces, including metals, rectifying semiconductors, and graphene, to produce wet photoelectrochemical cells. Specifically, I will describe how monolayers of PSI can be deposited onto electrodes using self-assembly and Langmuir-Blodgett/Schaefer methods as well as how multilayers of PSI are deposited through a vacuum-assisted drop-casting approach. Thicker films of PSI absorb more light and yield much greater photocurrents as compared to monolayer films. The electrode that these films are deposited on is vitally important for the performance of the film. We have observed that micron-thick PSI films on p-doped silicon produce 200-times greater photocurrents than those obtained by deposition on gold. The position of the Fermi level combined with the band gap of silicon enables unidirectional electron transfer to PSI so that we avoid current cancellation from non-optimally aligned proteins. Adsorption of PSI on graphene provides flexibility in cell design due to the conductive, transparent nature of the atomically thin electrode. Furthermore, integrating a PSI film with reduced graphene oxide improves film conductivity and enhances observed photocurrents. These discoveries are leading to vastly improved performances for PSI-biohybrid solar cells.

(262d) Exciton Diffusion in Quantum-Dot Thin Films

Tisdale, W. A., MIT

Originally presented on: 11/5/2013 9:27:00 - 9:46:00

Colloidal quantum dots (QD), also known as semiconductor nanocrystals, are an intriguing material platform for solution-processable optoelectronic devices, such as solar cells, light-emitting diodes, thermoelectric modules, and flexible electronics. In QD thin films, the efficient movement of charge, heat, and excitonic energy is intimately linked to the QD surface chemistry and the amount of structural or energetic disorder in the material. In excitonic solar cells, the efficient diffusion of excitons to charge-separating interfaces is central to device operation. On the other hand, exciton diffusion to quenching interfaces in light-emitting diodes is a process that can limit luminescence efficiency. While the diffusion of singlet and triplet excitons in organic semiconductors has been studied extensively, exciton diffusion in colloidal QD thin films remains largely unexplored. In this talk, I will detail my group's combined experimental and computational efforts to obtain a deeper understanding of excitonic energy transport phenomena in colloidal QD materials. These efforts include spectrally-resolved transient photoluminescence spectroscopy, photoluminescence quenching, time-resolved optical imaging, and kinetic Monte Carlo simulation.

(262e) Photonic Curing of Nanocrystals for Photovoltaics

Stolle, C. J., The University of Texas at Austin
Harvey, T. B., The University of Texas at Austin
Pernik, D. R., The University of Texas at Austin
Du, J., The University of Texas at Austin
Rhee, D.,
Korgel, B. A., The University of Texas at Austin
Originally presented on: 11/5/2013 9:46:00 - 10:05:00

Nanocrystal inks are being explored as a route to low cost high efficiency photovoltaic (PV) devices. The efficiency of nanocrystal devices is limited by charge transport through the nanocrystal film and higher efficiencies can be achieved by sintering the nanocrystals into polycrystalline films. For copper-indium-selenide (CIS), this is done by thermal annealing under a selenium rich atmosphere above 500°C (called selenization). This processing step is relatively expensive and difficult to control.

Therefore, we have been exploring nanocrystal film processing strategies that might alleviate this step. Here, we present results on the use of rapid photonic curing using a rapid broadband pulsed light source to transiently heat the sample, which allows for sintering of CIS films on metallic contacts and helps prevent the outgassing of selenium from the film. We demonstrate a working solar cell fabricated from a photonically sintered CIS film. The choice of back contact material is critical for achieving working sintered devices, as the thermal conductivity and optical properties of the contact have a large impact on the film heating. We also demonstrate the removal of organic capping ligand from the film without inducing grain growth. In this case, nanocrystals come into direct contact with each other and charge transport is greatly increased. Devices made from nanocrystal films without capping ligands demonstrate high short circuit current and external quantum efficiency, although the overall device efficiency is limited due to a low open circuit voltage.

(262f) Chloride Surface Modified Cadmium Telluride Nanocrystals for Photovoltaics

Hellebusch, D. J., University of California, Berkeley
Alivisatos, A. P., Lawrence Berkeley National Laboratory
Originally presented on: 11/5/2013 10:05:00 - 10:24:00

Semiconductor nanocrystals (NCs) are of great scientific and commercial interest due to their optoelectronic properties. However, improvements in nanocrystals for solar cell applications, specifically cadmium telluride, require the ability to controllably manipulate their surface. Halides are ideal for nanocrystal surface functionalization as their small size provides advantages in passivation, coupling, and processing compared to the ubiquitous long-chained aliphatic surfactants. Here, we present the controlled surface modification of cadmium telluride nanocrystals with chloride while maintaining colloidal stability in organic solvents with a secondary labile ligand. After deposition of NCs from solution, the secondary ligand is easily removed with vacuum leaving behind an entirely organic-free NC film, offering a processing advantage for solution-based semiconductor thin films. X-ray Photoelectron Spectroscopy, Rutherford Backscatter Spectroscopy, and H-NMR supports successful substitution of the native ligand with the halide species while UV-Vis Absorption Spectroscopy and Transmission Electron Microscopy confirm maintenance of particle size, shape and solubility in organic solvents. Furthermore, we investigate the effect of the chloride on thermally treated, or sintered, NCs films. X-Ray Diffraction and Small Angle X-Ray Scattering reveal recrystallization and grain growth while Capacitance-Voltage, Field Effect Transistors, and Time-Resolved Photoluminescence probe enhancements in electrical and optical properties of sintered NC films. Our material system offers a unique method to deposit films from solution with easily removable secondary ligands that tend to contaminate films with electrically insulating materials. This study introduces a nanocrystal system with a quantitative handle on surface manipulations for a broad range of investigations and applications for photovoltaics.

(262g) Multistep Selenization of Copper Indium Gallium Selenide (CIGS) Nanocrystal Photovoltaics

Harvey, T. B., The University of Texas at Austin
Bogart, T., The University of Texas at Austin
Stolle, C. J., The University of Texas at Austin
Du, J., The University of Texas at Austin
Pernik, D. R., The University of Texas at Austin
Korgel, B. A., The University of Texas at Austin
Originally presented on: 11/5/2013 10:24:00 - 10:43:00

Copper Indium Gallium Selenide (CIGS) nanocrystal inks have been used to fabricate photovoltaic devices without post-deposition thermal treatment, but have been limited to relatively low efficiency (~3%) by poor electron transport. The insulating organic nanocrystal capping ligands are largely responsible for this. Annealing the nanocrystal films at high temperature under selenium atmosphere—i.e., selenization—can be used to improve device performance by removing ligands and increasing grain size in the film. The selenization process, however, is difficult to control due to non-uniform film sintering, the formation of insulating layers from residual carbon in the film, and film delamination with complete conversion of the Mo back contact to $MoSe_2$.

We have found that a one hour thermal anneal under argon at 525 °C before selenization of the nanocrystal film can significantly improve device efficiency. This pre-selenization anneal removes organic ligands, which appears to be important for uniform nanocrystal sintering. Lowering the anneal temperature from 525 °C to 425 °C drastically lowers device efficiency although the capping ligands are still removed. Sodium has been shown to improve CIGS grain growth in co-evaporated films and a similar connection is found in selenized nanocrystal film performance. X-ray photoelectron spectroscopy (XPS) measurements taken after the pre-selenization anneal show increased Na presence at the Mo surface when the annealing temperature is increased due to greater diffusion from the soda-lime glass substrate.

Additionally, this pre-selenization anneal accelerates grain growth in the film, shortening the necessary selenization time. Shortened selenization length reduces $MoSe_2$ formation at the back contact, enabling the use of repeated selenization treatments without film delamination. Devices with over 7% efficiency have been fabricated by depositing and selenizing successive CIGS nanocrystal layers.

(262h) Exciton Dissociation and Charge Carrier Generation At Core/Shell Heterojunction in Quantum Dots

Chakraborty, A., Syracuse University
Originally presented on: 11/5/2013 10:43:00 - 11:02:00

Electron-hole pairs (excitons) are generated by photoexcitation of electrons to excited electronic states. Exciton dissociation and generation of free charge carriers is central for light-harvesting applications of photoactive nanoparticles. This talk will focus on computational investigation of effect of heterojunctions on exciton dissociation in core/shell quantum dots. It has been shown in various experimental studies that presence of heterojunctions in nanocrystals can greatly facilitate exciton dissociation. This study will present the following four computed properties: (1) exciton binding energy, (2). electron-hole recombination probability (3). electron-hole separation distance and (4).electron-hole pair density as metrics for analyzing effect of core/shell structures in nanocrystals. Results for CdSe/ZnS core/shell quantum dots will be presented for a series of dots sizes. Effect of dot size and thickness of the shell of exciton dissociation will be presented and comparison with experimental results will be discussed. The electron-hole interaction energy is quantified by calculating the exciton binding energy and the spatial distribution of the charge carrier in the quantum dot is analyzed by electron-hole pair density functions. The local environment of the electron and hole quasiparticles in the quantum dot is probed using the average electron-hole separation distance and the electron-hole recombination probability. Exciton dissociation is a complex process and the four metrics mentioned above provide a multi-faceted description of this phenomenon.

(318a) Smaller Than the Bandgap: The Tale of Low-Energy Photons

Sun, Q. C.,
Lien, C. H.,
Nagpal, P., University of Colorado Boulder
Originally presented on: 11/5/2013 12:30:00 - 12:49:00

Sun emits light in a broad spectrum of electromagnetic radiation. Conversion of this wide spectrum of incident radiation into more useful forms of energy, like electricity, using finite bandgap semiconductors leads to inefficiency. Since the photons with energy smaller than the semiconductor bandgap are simply transmitted through the photocell, more than 30% of incident radiation is not absorbed in the best photovoltaic cells available now. Utilizing this near-infrared, sub-bandgap radiation requires application of novel physical phenomena like upconversion, energy-transfer, and generation of low-energy hybrid photon-electron waves known as surface plasmon polaritons. I will summarize the efforts in my group towards utilization of sub-bandgap photons by a combination of single nanoparticle spectroscopy to optimize these novel physical phenomena, and translation of these concepts into device scale optoelectronics. These photonic devices can be combined with a wide-variety of semiconductor photocells for improving photovoltaic conversion efficiencies. These results can have important implications for solar energy conversion, and for developing a new class of bio-nanomaterials for optics-based sensing in the region of optical transparency for biological tissues.

(318b) Polymer-Free Near-Infrared Photovoltaics With Single Chirality (6,5) Semiconducting Carbon Nanotube Active Layers

Jain, R., Massachusetts Institute of Technology
Howden, R., Massachusetts Institute of Technology
Tvrdy, K., Massachusetts Institute of Technology
Shimizu, S., Massachusetts Institute of Technology
Hilmer, A. J., Massachusetts Institute of Technology
McNicholas, T. P., Massachusetts Institute of Technology
Gleason, K. K., Massachusetts Insitute of Technology
Strano, M. S., Massachusetts Institute of Technology
Originally presented on: 11/5/2013 12:49:00 - 13:08:00

The addition of single-walled carbon nanototubes (SWNTs) into next generation solar cells as either near infrared absorbers or directional charge transporters has shown the potential to increase device efficiencies. Previous measurements of single and few tube devices implicate nanotube-nanotube junctions in exciton recombination, limiting efficiency. Here, we ask if this limitation scales to an all nanotube hole conducting phase, or the unity volume fraction limit of an all carbon device. We create such a (6,5) SWNT/C60 photovoltaic device for the first time. Devices yielded 0.5% external quantum efficiency (EQE) at the peak of the nanotube absorption spectrum and a power conversion efficiency of 0.10% under AM1.5 illumination. Interestingly, this value exceeds many of those reported for polymer/SWNT/PCBM composite based photovoltaics. Further, the addition of only 20% (6,4) SWNTs to an otherwise pure (6,5) SWNT film reduces device efficiency 30 times, demonstrating the importance of an energetically homogeneous film with respect to SWNT chirality. Through the construction of an all-carbon photovoltaic, we show both the feasibility of this novel device as well as inform the mechanisms for inefficiencies in SWNT based solar cells.

(318c) Suppressing Optical Absorption in Nanostructured Metal Electrodes in Photovoltaics

Han, S. E.,

Originally presented on: 11/5/2013 13:08:00 - 13:27:00

Electrodes are ubiquitously used in optoelectronic devices such as solar cells, infrared detectors, and light emitting diodes. Typically, metals exhibit good electrical conductivity due to their free electrons and are suitable for electrodes in these devices. However, metal electrodes cause inefficiency in the devices by absorbing light as their free electrons suffer from collisions. To circumvent this problem, nanostructured metals are being explored to realize low optical losses while maintaining large electrical conductivity. Here, we study optical absorption in helical metal nanocoils and find that absorption can be dramatically decreased when the metal is suitably nanostructured. Theoretical modeling showed that this effect is due to the increase in effective mass of free electrons in nanostructured metals. Heavy electrons suffer a decreased rate of collision and emulate dielectric materials where optical losses are negligible. Calculations showed that, when metal electrodes are used in semiconductors devices, suitably chosen nanostructures can increase the fractional absorption in semiconductors more than 30 times by reducing losses in the metal.

(318d) Nanopore-Type Black Silicon Anti-Reflection Layers Fabricated By a One-Step Silver-Assisted Chemical Etching

Lu, Y. T., Rice University
Barron, A. R., Rice University
Originally presented on: 11/5/2013 13:27:00 - 13:46:00

Solar cells need low surface reflectance to maximize the amount of absorbed incident photons for improving the energy conversion efficiency. Anti-reflection (AR) coatings are traditional materials to suppress the reflection of silicon (Si)-based solar cell surface, but can only work for a narrow range of light wavelength and incident angle. To overcome the disadvantages of AR coatings, a novel type of porous Si material called black silicon (b-Si) has been considered as replacement for the AR coatings. B-Si is directly formed from the Si wafer of the solar cell and its surface possesses a graded refractive index resulting in a low reflectivity of visible light. Although b-Si has been successfully fabricated by several different methods, most of the techniques need either expensive instruments and high energy consumption or complicated fabricating processes, causing them unfavorable for industrial applications.

Recently, an economical and efficient approach for industrial b-Si manufacturing called metal-assisted chemical etching method has been developed to avoid the above disadvantages. The metal-assisted chemical etching method includes two steps: metal deposition and electroless chemical etching. In the metal deposition step, a noble metal, such as Au, Ag, and Pt is deposited on the Si surface usually as nanoparticles (NPs). The metal NPs attract electrons from the silicon surface promoting the oxidation to SiO_2 in the presence of an appropriate oxidant. In the electroless chemical etching step, the as-formed SiO_2 is etched away by HF (as H_2SiF_6) and a pit is produced under each NP. As these reactions occur in a continuous process the pits become deeper and ultimately connect with each other, and remaining Si substrate forms b-Si that consists of a silicon nanowire (Si-NW) structure. Unfortunately, this Si-NW structure is very fragile making the incorporation of b-Si into typical fab processes difficult.

The proposed method here, one-step silver (Ag)-assisted chemical etching method, is more facile technique which can synthesize b-Si in 20 minutes under room temperature without any pricy facility. During the process of the one-step method, the deposition of Ag NPs and the chemical etching of Si occurred simultaneously on the Si wafer surface. In our b-Si fabrication process, cleaned Si wafers were etched in a Si etchant consist of HF, H_2O_2, H_2O, and $AgNO_3$ with various concentrations in sealed plastic ware at room temperature.The effects on the surface morphology and the corresponding surface reflectivity of the concentration of the Ag catalyst precursor, the HF and H_2O_2 concentrations in the Si etchant, and etching time have been systematically investigated. The $AgNO_3$ was utilized as an Ag NP precursor in a $HF:H_2O_2:H_2O$ solution to compose of a Si etchant; the H_2O_2 not only facilitates the Si etching but also reduces Ag^+ ions to Ag NPs on the Si wafer. After the Si etching, the Ag remained on the b-Si samples were removed by an Ag etchant consisting of NH_4OH and H_2O_2. Compared to other methods, the one-step Ag-assisted chemical etching simplifies the b-Si fabrication process and may cut down the cost of the facilities and energy expenditures, which is beneficial for industrial applications. In addition, unlike Si-NW arrays, the nanopore-type b-Si possesses no high-aspect-ratio needle-liked structure and is not as fragile as Si-NW arrays and can thus better endure the stress during the solar cell assembly.

According to the results, we found that lower reflectivity is a balance between sufficient Ag catalyst to create large numbers of nanopores on a Si surface and excessive Ag that brings deeply etched channels which would potentially short-circuit a solar cell junction. In addition, a mechanism for describing the formation process of black Si by one-step Ag-assisted chemical etching method is also proposed. With $[Ag^+]$ of 500 μM, nanopores are hardly observed on the Si wafer surface causing the relative reflectivity of Si wafer surface cannot be effectively improved. The lowest average relative reflectivity, 0.17%, occurs with a $[Ag^+]$ of 50 μM and the $HF:H_2O_2:H_2O$ ratio of 1:5:2, with nanopore length longer than 6 μm. However, as the $[Ag^+]$ decreases to 5 μM, a low relative reflectivity (2.60%) and a short nanopore length (< 250 nm) can be obtained by 10 minute etching time and the $HF:H_2O_2:H_2O$ ratio of 5:1:20, indicating that this method can be used as a simple (one-pot), low cost (low $[Ag^+]$), energy efficient (room temperature), method for the synthesis of AR layers for Si-based solar cell applications.

We propose that the simplicity of the process and low reaction times, along low consumption of Ag precursor will make such a one-step method scalable for industrial AR applications where process cost and rate are important. Moreover, the nanopore length of b-Si fabricated with 5 μM $[Ag^+]$ can be controlled to be around 250 nm, which is shorter than a typical p-n junction depth in Si-based solar cells, and would not interfere with the function of the cell.

(318e) Band Alignment of Ternary Lead Chalcogenide Quantum Dots Via Composition and Size Tuning for Photovoltaic Applications

Palmstrom, A.,
Santra, P., Stanford University
Bent, S. F., Stanford University
Originally presented on: 11/5/2013 13:46:00 - 14:05:00

Lead chalcogenide quantum dots have recently been established as a strong candidate for quantum dot sensitized solar cells due to a low, tunable band gap and a large absorption coefficient. The band gap of bulk PbS and PbSe is 0.41 eV and 0.28 eV with large Bohr excitonic radius of 18 nm and 46 nm respectively. These band gaps can be increased through size quantization. Here we report the synthesis of ternary lead chalogenide ($PbSe_xS_{1-x}$) quantum dots by hot injection synthesis at temperatures below 150 °C that absorb in the range of 900 nm to 1200 nm. The band gaps of these quantum dots are tuned by changing the ratio of sulfur to selenium while maintaining a constant size. Furthermore, the effects of passivating ligands, injection temperature, as well as the growth time of these quantum dots were investigated to achieve better control over the size, band gap, and crystallinity of the particles. Ultraviolet photoemission spectroscopy and optical absorption spectroscopy were used to determine the exact positions of the valence band maxima and conduction band minima in order to improve the design of quantum dot solar cells.

(318f) Charge Transfer Structure-Reactivity Dependence of Fullerene/Single-Walled Carbon Nanotube Heterojunctions

Hilmer, A. J., Massachusetts Institute of Technology
Tvrdy, K., Massachusetts Institute of Technology
Zhang, J., Massachusetts Institute of Technology
Strano, M. S., Massachusetts Institute of Technology
Originally presented on: 11/5/2013 14:05:00 - 14:24:00

Charge transfer at the interface between single-walled carbon nanotubes of distinct chiral vectors and fullerenes of various molecular weights is of interest both fundamentally and because of its importance in emerging photovoltaic and optoelectronic devices. One approach for generating isolated, discretized fullerene/SWCNT heterojunctions for spectroscopic investigation is to form an amphiphile from the former to disperse the latter at the single-SWCNT level in aqueous solution. Herein, we synthesize a series of methanofullerene amphiphiles, including derivatives of C_{60}, C_{70}, and C_{84}, and investigated their electron transfer with SWCNT of specific chirality, generating a structure/reactivity relationship. For C_{61} and C_{71} methanofullerenes, the driving force for photoexcited electron transfer from SWCNT is small, and therefore significant decreases in the relative SWCNT fluorescence yield are only observed in the limit of high surface coverage of the methanofullerene moiety. In the case of 29% and 26% surface coverage by the C_{61} and C_{71} methanofullerenes, respectively, only a partial quenching of nanotube fluorescence was observed across all species. In the case of C_{61} it was observed that near-complete quenching could be obtained by increasing the surface coverage of C_{61} to 92%. In contrast, in the case of C_{85} methanofullerene, even at a predicted SWCNT surface coverage of only 14%, this fullerene derivative shows complete quenching of all semiconducting SWCNT utilized in this work. This enhancement in quenching efficiency is consistent with the deeper LUMO level of C_{85} methanofullerene over comparatively smaller fullerene adducts, and suggests its promise as for SWCNT/fullerene heterojunctions. The relative rates of charge transfer for C_{61} and C_{71} complexes with SWCNT of varying bandgap are well described by a Marcus theory based structure-reactivity relationship.

(318g) Nano- and Molecular Precursors for CdTe Photovoltaics

Panthani, M. G., University of Chicago
Dolzhnikov, D., University of Chicago
Dietz, T., University of Chicago
Crisp, R., National Renewable Energy Laboratory
Luther, J.,
Talapin, D. V., University of Chicago
Originally presented on: 11/5/2013 14:24:00 - 14:43:00

Solution-processable semiconductors offer the potential for inexpensive and high-throughput processing of photovoltaics with high materials utilization. One approach that has been used to successfully produce high efficiency solar cells is to use solution-deposited materials as a nanocrystalline or molecular semiconducting thin film that can be annealed into a large-grained material. We report approaches to producing crystalline CdTe films using nanocrystals capped with either pyridine or metal-chalcogenide complexes (MCCs) as well as MCCs alone. We find that with low temperature (~350°C) sintering in air, the grain size can be increased from ~5 nm to over 100 nm. Out best solar cells show high short circuit currents (~24 mA cm^{-2}) but low open circuit voltages (~600 - 700 mV). Additionally, the devices require current and/or light soaking to achieve their maximum efficiency. In optimized devices, power conversion efficiencies of up to 9.7% have been achieved in a ITO/CdTe/ZnO/Al configuration. We find that the efficiency may be limited by non-ideal contacts and offer some insight into ways the open circuit voltage could be improved.

(318h) Shape-Selective Formation of Ag Nanocrystals: Insights From Atomic-Scale Simulations

Fichthorn, K., The Pennsylvania State University
Zhou, Y., Pennsylvania State University
Originally presented on: 11/5/2013 14:43:00 - 15:02:00

The ability of plasmonic Ag and Au nanostructures to concentrate UV-vis radiation in small volumes makes it attractive to use these building blocks in the design of composite photocatalysts for the production of solar fuels, as nanometallic plasmonic structures to enhance the efficiency of photovoltaics, or as flexible electrode materials. A popular method of synthesizing Au and Ag nanocrystals is the solution-phase polyol process, in which a structure-directing agent (commonly PVP) directs the growth of seed crystals to form shape-selective nanostructures. The exact role of PVP in this synthesis is unclear. It is likely that PVP protects growing Ag nanocrystals from uncontrolled aggregation. PVP also promotes the formation of {100}-faceted Ag nanostructures in solution, while the {111} facet has the minimum energy in vacuum. The reasons for this facet selection are unclear and raises the question: Does PVP affect the fluxes of solution-phase Ag atoms to promote growth of nonequilibrium structures with {100} facets? Or does PVP simply alter the solution-phase interfacial energies of the two facets, so that the {100} facet is thermodynamically preferred in solution? To answer these questions, we used molecular dynamics simulations based on force fields fitted to first-principles density-functional theory calculations to calculate the interfacial energies of PVP-covered Ag surfaces in ethylene glycol solvent. In this way, we mimic the solution-phase environment that occurs for Ag nanostructure synthesis. We investigate the interfacial energies of PVP-covered Ag(100) and Ag(111). Additionally, we calculate the flux of Ag atoms to Ag surfaces that are both bare (in solvent) and covered with PVP. We find that the flux of Ag atoms to PVP-covered Ag(111) is larger than to Ag(100), which would promote the growth of Ag(100) facets in a crystal. Interestingly, the flux of Ag atoms to PVP-covered surfaces is larger than that to bare surfaces in solvent. This suggests a non-equilibrium growth scenario, in which PVP channels atoms selectively to {111} facets to promote the growth of {100}-faceted nanostructures.

(319a) Core-Shell Structured Composite Microparticles With Ability to Store a Chemical Payload and Release a Defined Quantity "On Demand"

Kovacik, P., Prague Institute of Chemical Technology
Stepanek, F., Institute of Chemical Technology, Prague
Singh, M.,
Originally presented on: 11/5/2013 12:30:00 - 12:55:00

Core-shell structured composite microparticles with ability to store a chemical payload and release a defined quantity ?on demand?

P. Kovacik, M. Singh, F.Stepanek

ICT Prague, Laboratory of chemical robotics, Technicka 5, 166 28 Prague,
Tel.: 2 2044 3046, e-mail: kovacikp@vscht.cz

The aim of this work was the preparation of biocompatible microparticles with a hollow interrior. The microparticles must be able to encapsulate and store a chemical payload for a certain time, followed by the release on demand of this payload. In our previous work [1] we have prepared silica microparticles with a hollow core and found that the diffusion across the mesoporous silica shell was strongly dependent on temperature. In this work, we used this dependence and attached iron oxide nanoparticles on the surface of the silica shell to create composite iron oxide/silica microparticles. The iron oxide nanoparticles were able to heat up in the presence of an alternating magnetic field. This property allowed us to use magnetic field as a tool for remote control of diffusion across the microparticle shell. To avoid spontaneous leakage of encapsulated payload in time, we have modified the surface of the composite microparticles with a layer of palm oil. Palm oil is a phase change material which is solid under 37 Â°C. We showed that the resultant composite microparticles are able to store a payload for several months and release a defined quantity on demand by the application of a magnetic field. The particles were characterised in shape, size, heating ability and their mass transport properties.

[1] Kovacik P., Kremlackova Z., Stepanek F., Investigation of radiofrequency induced release kinetics from magnetic hollow silica microspheres, Micropor. Mesopor. Mater. 159, 119-125 (2012)

(319b) ZnO Nanorods Hermetically Encapsulated By a Nanothin Amorphous SiO2 Coating: Toxicological Profile and Optical Properties

Sotiriou, G. A., Harvard University
Murdaugh, K.,
Cohen, J., Harvard University, School of Public Health
Watson, C.,
Elder, A.,
Demokritou, P., Harvard School of Public Health
Originally presented on: 11/5/2013 12:55:00 - 13:20:00

Zinc oxide (ZnO) is a wide-bandgap metal oxide semiconductor that can be excited at room temperature by UV irradiation, and finds applications in a variety of fields and products including paints, pigments, batteries, photocatalysis, foods, to name a few. When in nanometer size range, it may be used in polymer nanocomposites and sunscreens as an efficient UV-filter with high transparency in the visible wavelength range. However, the photocatalytic activity of ZnO causes the degradation of the surrounding polymer rendering it unsuitable for long-term employment. Furthermore, ZnO nanoparticles are highly toxic and may pose risks to the public and the environment. Here, ZnO nanorods are made by scalable flame synthesis [1,2] and are in-situ encapsulated by an amorphous nanothin SiO_2 layer [3,4]. The as-prepared nanoparticles are characterized by electron microscopy (EM), X-ray diffraction (XRD), X-ray photoelectron spectroscopy (XPS) and N_2 adsorption. The hermetic nature of the SiO_2 coating is evaluated by a detailed dissolution study and the photocatalytic activity is monitored by the decomposition of methylene blue dye. The presence of SiO_2 facilitates the dispersion of these nanorods in relevant solutions. The core ZnO nanorods exhibit the characteristic optical properties as determined by UV/vis and luminescent spectrometry, while the SiO_2 coating eliminates the photocatalytic activity and minimizes the toxicity and genotoxicity to human cells, as illustrated in cellular studies using multiple cell lines. Therefore, these flame-made ZnO nanorods may be safely used as fillers in polymer UV-filter nanocomposites and in sunscreens exhibiting superior performance while mitigating their impact on environmental health.

References

[1] Sotiriou, G. A., Sannomiya, T., Teleki, A., Krumeich, F., Vörös, J. and Pratsinis, S. E. Non-toxic dry-coated nanosilver for plasmonic biosensors. *Adv. Funct. Mater.* **20**, 4250-4257, (2010).

[2] Sotiriou, G. A., Hirt, A. M., Lozach, P. Y., Teleki, A., Krumeich, F. and Pratsinis, S. E. Hybrid, silica-coated, Janus-like plasmonic-magnetic nanoparticles. *Chem. Mater.* **23**, 1985-1992, (2011).

[3] Sotiriou, G. A., Franco, D., Poulikakos, D. and Ferrari, A. Optically Stable Biocompatible Flame-Made SiO_2-Coated Y_2O_3:Tb^{3+} Nanophosphors for Cell Imaging. *ACS Nano* **6**, 3888-3897, (2012).

[4] Gass, S., Cohen, J. M., Pyrgiotakis, G., Sotiriou, G. A., Pratsinis, S. E. and Demokritou, P. Safer Formulation Concept for Flame-Generated Engineered Nanomaterials. *ACS Sustainable Chem. Eng.* **in press**, DOI: 10.1021/sc300152f, (2013).

(319c) Single-Pot Synthesis of Uniform Glucan Multilayers On Oxide Particles

Jankolovits, J.,
Gazit, O., University of California Berkeley
Nigra, M., University of California, Berkeley
Katz, A., University of California, Berkeley
Originally presented on: 11/5/2013 13:20:00 - 13:45:00

Title: Single-Pot Synthesis of Uniform Glucan Multilayers on Oxide Particles

Authors: Joseph Jankolovits, Oz M. Gazit, Michael M. Nigra, Alexander Katz

Department of Chemical and Biomolecular Engineering

University of California, Berkeley

Polymer coatings are used widely to tune the surface chemistry of solids in biomaterials, paints, and other materials. Covalently grafted polymer coatings provide advantageous stability, though their prevailing syntheses involve multiple step procedures, low solid yields and surface coverage, and/or non-uniformity on the nanoscale. A novel method for uniformly grafting polymer multilayers on inorganic oxide particles in a single-pot process will be demonstrated using a wide variety of characterization techniques. Implications of the grafted polymer multilayer coatings on inorganic oxide particles for the stabilization of paint pigment dispersions will be discussed. Efforts to prepare grafted polymer multilayers on nanoparticles will be described. The promise of these grafted polymer multilayer coatings on inorganic oxide particles in biomaterials and nanomedicine will be highlighted.

(319d) Continuous Polymer Coating of Nanoparticles: A Novel Method

Sirkar, K. K., New Jersey Institute of Technology
Chen, D., New Jersey Institute of Technology
Singh, D., New Jersey Institute of Technology
Pfeffer, R., Arizona State University
Originally presented on: 11/5/2013 13:45:00 - 14:10:00

Nanoparticle-based drug delivery systems are of significant interest in controlled release of drugs, delivery of anticancer drugs and imaging agents to tumors, tuberculosis treatment and as non-viral gene delivery vehicles. To develop controlled release of the drug as well as to protect the drug from demanding environments, appropriate coatings are employed on the drug nanoparticles. Conventional methods for coating or encapsulating micron–sized and nanoparticles utilize dry or wet approaches. Dry methods include physical vapor deposition, plasma treatment, chemical vapor deposition and pyrolysis of polymeric organic materials. Wet methods cover sol-gel processes, emulsification and solvent evaporation techniques. Supercritical fluid (SCF) processes such as Rapid Expansion of Supercritical Solutions (RESS), Supercritical Anti-Solvent (SAS), and Gas Anti-Solvent (GAS) processes employing supercritical CO_2 are alternative methods for nanoparticle coating or encapsulation of ultrafine particles. These SCF-based processes have many shortcomings. Some processes require demanding operating conditions, suffer from low polymer solubility etc. Most of these techniques are also batch processes. Fluidized bed-based processes, which can be continuous, face difficulties with nanoparticles caused by van der Waals and other interparticle forces. In all such processes, scale-up is also quite demanding. We will illustrate a technique for continuous polymer coating of nanoparticles that is easily scalable. The nanoparticles used were of silica. Polymers used for coating were Eudragit and PLGA.

(319e) Coating FINE Particles With ULTRA Thin FILMS Using Atomic Layer Deposition

Meesters, G., DSM
Valdesueiro, D., Delft University of Technology
van Ommen, J. R., Delft University of Technology
Originally presented on: 11/5/2013 14:10:00 - 14:35:00

Atomic Layer Deposition (ALD) is a gas-phase coating technique that can provide fine particles with ultrathin films, typically 1-50 nm. Due to the large value of the specific surface area, fine particles –diameter ranging from 1 nm to 50 μm– are used in a wide range of industrial fields such as electronics, pharmaceutics, energy storage and catalysis. The aim of the coating is to improve the performance of these materials, either by the activation of the surface of the particles, or protecting them against the environment.

ALD can deposit uniform inorganic layers [1] by a set of two self-limiting reactions (Fig. 1). The precursors, which are fed into the reactor in subsequent steps, react on the surface of the particles to form a very thin layer of coating material. The final thickness of the coating is controlled by repeating the cycle a desired number of times. This layer-by-layer growth mechanism can also be used to deposit polymeric compounds (Molecular Layer Deposition) [2] or inorganic-organic hybrid compounds [3]. The coating process is carried out in a fluidized bed reactor, which offers high mass and energy transfer, and the possibility to process large amounts of solids.

The production of core-shell particles via ALD has been already applied in fields such as the Li-ion batteries [1] (Fig. 2) or photo-catalysis [4]. Our current work aims at the production of non-permanent coatings to be applied in the controlled release of compounds for food and pharmaceutical applications.

Fig. 1. Schematic representation of the two ALD half-reactions.

Fig. 2. Li-ion cathode material (LiMn2O4) coated with 5 cycles of Al2O3 using ALD [1]

1. Van Ommen, J.R., et al., *Scalable gas-phase processes to create nanostructured particles.* Particuology, 2010. **8**(6): p. 572-577.

2. Adamczyk, N.M., A.A. Dameron, and S.M. George, *Molecular layer deposition of poly(p-phenylene terephthalamide) films using terephthaloyl chloride and p-phenylenediamine.* Langmuir, 2008. **24**(5): p. 2081-2089.

3. Yoon, B., et al., *Molecular layer deposition of hybrid organic - Inorganic alucone polymer films using a three-step abc reaction sequence.* Chemistry of Materials, 2009. **21**(22): p. 5365-5374.

4. Zhou, Y., et al., *Optimal preparation of Pt/TiO2 photocatalysts using atomic layer deposition.* Applied Catalysis B: Environmental, 2010. **101**(1-2): p. 54-60.

(319f) Surface Engineered Quantum Dots for Light Selective Polymer Films

Mumin, M. A., University of Western Ontario
Allan, J. M., Western University
Charpentier, P. A., Western University
Originally presented on: 11/5/2013 14:35:00 - 15:00:00

We have investigated the optimization of the optical properties of polymer films integrating engineered quantum dot (QD) nanocrystals to manipulate and control light transmission for applications in light selective polymer films. Highly luminescent CdS QDs and CdS/ZnS core/shell QDs with 5 nm size were synthesized using a facile approach based on the pyrolysis of single molecule precursors. After capping the CdS QDs with a thin layer of ZnS to reduce toxicity, the photoluminescence and photostability of the core-shell QDs was significantly enhanced. The ZnS shell was found to passivate the surface defects of the CdS dots while confining the electron-hole pair to the CdS core region.

To make both the bare and core/shell structure QDs more resistant against photochemical reactions, a mesoporous silica layer was grown on the QDs through a reverse microemulsion technique based on hydrophobic interaction and surface silanization. Mesoporous silica can protect completely both type of QDs from photobleaching and also improves the quantum yield by providing much stronger resistance to oxidation and Oswald ripening of QDs. The encapsulated QDs were then functionalized using silane coupling agent and integrated to poly ethylene vinyl acetate (pEVA) network both by melt mixing (using twin screw extruder) and grafting from polymerization in supercritical CO_2. Morphology, UV absorption, visible light transmission, and IR retention were observed and compared to that obtained from commercial pEVA greenhouse plastics. Mesoporous silica encapsulated QDs into EVA polymer films were found to provide simultaneous UV protection, infrared retention, and light selectivity of solar-polymer films.

(374a) Simulation Tools for Nanoparticle-Based Composite Processing and Property Prediction

Bolintineanu, D. S., University of Minnesota
Lechman, J. B., Sandia National Laboratories
Schunk, P. R., Sandia National Laboratories
Originally presented on: 11/5/2013 15:15:00 - 15:40:00

We present a set of tools for the modeling of various aspects of solution-based nanomanufacturing processes. First, we discuss several discrete element modeling (DEM) techniques for the simulation of nanoparticle/colloidal suspensions, including efficient algorithms for the inclusion of hydrodynamic interactions, capillary forces, long-range colloidal interparticle forces, granular contact forces, solvent evaporation and highly aspherical particles. We apply these techniques to the simulation of a battery electrode manufacturing process, and show that the predicted microstructures provide relevant insights for the processing parameters. In the second half of the talk, we discuss methods for the prediction of the bulk properties of disordered, particle-based composites. We present several approaches that rely on a finite element model (FEM) representation of a sample of the material, which can be extracted from a DEM simulation or taken from a tomographic image of the real composite. Overall, this work presents an extensive, multiscale simulation toolset for connecting particle-based nanomanufacturing processing conditions to resulting microstructures and ultimately to material properties.

(374b) Tin Oxide Nanowires and Their Hybrid Architectures for Kinetically Fast Redox Couples in Dye-Sensitized Solar Cells

Vendra, V. K., University of Louisville
Nguyen, T., University of Houston
Druffel, T., Conn Center for Renewable Energy Research
Sunkara, M., University of Louisville
Amos, D. A., University of Louisville
Originally presented on: 11/5/2013 15:40:00 - 16:05:00

The redox electrolyte is an important component of the dye-sensitized solar cells. Iodide/tri-iodide electrolyte used in high efficiency cell is not suitable for practical application due to its corrosive nature, toxicity, competition between dye and tri-iodide for light aborption. One of the major drawback of using alternate redox couples is their fast recombination kinetics and mass transport limitations.

In this study, we show that tin oxide nanowires show a substantial improvement in the photovoltaic performance when compared to titania nanoparticles, when electrolytes having fast recombination kinetics are employed in dye-sensitized solar cells (DSCs). State of the art methods in literature employ titania nanoparticles use passivating layers of Al_2O_3 or poly(methyl siloxane) to lower the electron recombination rate. There have been reports investigating the performance of nanowire in conjunction with alternate redox couples. The single crystalline nature allows for fast electron transport in tin oxide nanowires. Further, prior work from our group has also shown that tin oxide nanowires exhibit two orders of magnitude higher electron lifetimes when compared to titania nanoparticles when iodide/tri-iodide is used as an electrolyte. In this study we employed tin oxide nanowires as photoanodes for DSCs and investigated their performance when kinetically fast electrolytes such as ferrocene/ferrocenium,tempo/tempo,ferricyanide /ferrocyanide are used as redox electrolytes.

Surface modifications such as $TiCl_4$ treatment and atomic layer deposition of titania and alumina were performed to further lower electron recombination in tin oxide nanowires. Fundamental properties such as electron transport time scales and electron recombination time scales were determined for tin oxide nanowires and hybrid architectures where the surface of tin oxide nanowires is coated with titania and alumina.

(374c) Multiscale Sustainability Assessment in Nanocoating Material Design and Manufacturing

Song, H., Wayne State University
Uttarwar, R., Wayne State University
Huang, Y., Wayne State University
Originally presented on: 11/5/2013 16:05:00 - 16:30:00

Nanopaint has drawn great attention in the coatings industry, as it may offer surface coatings with significantly improved performance or new functionalities and properties, such as surface texture alteration control, self-healing, toxic chemical/acid/corrosive agent repelling, glass opaqueness control, building exterior's infrared reflecting/absorbing properties control, and even pathogen growth control. Nevertheless, it is under scrutiny for the environmental and health risks which may pose in nanopaint manufacturing and nanocoating realization. Therefore, it is crucial to analyze the sustainability performance of nanocoating materials and prepare a roadmap for their successful development and application in the future.

In this paper, we introduce a systematic sustainability assessment methodology by resorting to multiscale science and sustainability science. Different from known methodologies focusing on the assessment of manufacturing process and manufactured product, this methodology initiates the assessment starting at the nanomaterial formulation design stage. A variety of material design parameters, such as those about polymer system, nanomaterial type, shape, size, and orientation, solvent and additives, are analyzed using a number of sustainability indicators, which will be incorporated into the analysis of coating manufacturing stages. Multiscale models are developed to predict nanoparticle emission severity in nanopaint spray and curing, nanocoating properties, product realization process performance on quality assurance, waste reduction, and operational safety. The efficacy of the multiscale sustainability assessment methodology will be demonstrated by the case studies on nanocoatings with super-scratch resistance performance and barrier property. The introduced methodology is generic and can be applied to study the sustainability performance of a wide range of nanocoating materials in their early design stage.

(374d) Continuous Nanoparticle Sizing and Characterization Via Microfluidic Interfacial Fluorescent Complexation

Meng, F., Texas A&M University
Ugaz, V. M., Texas A&M University
Originally presented on: 11/5/2013 16:30:00 - 16:55:00

A variety of characterization tools are available to sensitively analyze collected nanoparticles (e.g., condensation particle counters, differential mobility analyzers, scanning mobility particle sizers (SMPS), ion chromatography, mass spectrometry, and electron microscopy). But these methods often require a dedicated laboratory infrastructure to operate optimally, and are therefore not straightforward to adapt for online use. We have developed a new microfluidic-based approach that overcomes many of these limitations by harnessing the inherently steep chemical gradient established at the interface between co-flowing streams containing a nanoparticle-laden suspension and a fluorescent dye. This sharp mismatch acts to localize adsorptive dye-nanoparticle complexation interactions within a narrow interfacial zone, instantaneously producing an intense and easily detectable fluorescence signature. These interactions are inherently dominated by phenomena occurring at the nanoparticle surface, introducing the exciting possibility to extract information about particle size and morphology from the fluorescence profile. We demonstrate this capability experimentally using ZnO and TiO_2 nanoparticles, and introduce a reaction model that enables interfacial fluorescence to be predicted so that kinetic parameters associated with the underlying surface complexation can be quantified. Our approach offers potential for continuous operation that is particularly attractive in manufacturing settings where robust on-line characterization tools are needed.

(374e) Safer Formulation Concept for Flame-Generated Engineered Nanomaterials

Sotiriou, G. A., ETH Zurich
Gass, S.,
Cohen, J., Harvard University, School of Public Health
Pyrgiotakis, G., University of Florida
Pratsinis, S. E., ETH Zurich
Demokritou, P., Harvard School of Public Health
Originally presented on: 11/5/2013 16:55:00 - 17:20:00

The likely success or failure of the nanotechnology industry depends on the environmental health and safety of engineered nanomaterials (ENMs). While efforts toward engineering safer ENMs are sparse, such efforts are considered crucial to the sustainability of the nanotech industry. A promising approach in this regard is to coat potentially toxic nanomaterials with a biologically inert layer of amorphous SiO_2. Core-shell particles exhibit the surface properties of their amorphous SiO_2 shell while maintaining specific functional properties of their core material. A major challenge in the development of functional core-shell particles is the design of scalable high-yield processes that can meet large-scale industrial demand. Here, we present a safer formulation concept for flame-generated ENMs based on a one-step, in flight SiO_2 encapsulation process, which was recently introduced by the authors as a means for a scalable manufacturing of SiO_2 coated ENMs. Firstly, the versatility of the SiO_2-coating process is demonstrated by applying it to four ENMs (CeO_2, ZnO, Fe_2O_3, Ag) marked by their prevalence in consumer products as well as their range in toxicity. The ENM-dependent coating fundamentals are assessed and process parameters are optimized for each ENM investigated. The effects of the SiO_2-coating on core material structure, composition and morphology, as well as the coating efficiency on each nanostructured material, are evaluated using state-of-the-art analytical methods (XRD, N_2 adsorption, TEM, XPS, isopropanol chemisorption). Finally, the biological interactions of SiO_2-coated vs. uncoated ENMs are evaluated using cellular and animal bioassays, providing valuable evidence for reduced toxicity for the SiO_2-coated ENMs. Results indicate that the proposed 'safer by design' concept bears great promise for scaled-up application in industry in order to reduce the toxicological profile of ENMs for certain applications.

References

[1] Sotiriou G.A., Sannomiya T., Teleki A., Krumeich F., Vörös J. and Pratsinis S.E. Non-toxic dry-coated nanosilver for plasmonic biosensors. *Adv. Funct. Mater.* **20**, 4250-4257, (2010).

[2] Sotiriou G.A., Hirt A.M., Lozach P.Y., Teleki A., Krumeich,F. and Pratsinis S.E. Hybrid, silica-coated, Janus-like plasmonic-magnetic nanoparticles. *Chem. Mater.* **23**, 1985-1992, (2011).

[3] Gass S., Cohen J.M., Pyrgiotakis G., Sotiriou G.A., Pratsinis S.E. and Demokritou P. Safer Formulation Concept for Flame-Generated Engineered Nanomaterials. *ACS Sustainable Chem. Eng.* **in press**, DOI: 10.1021/sc300152f, (2013).

[4] Demokritou P., Gass S., Pyrgiotakis G., Cohen J.M., Goldsmith W., McKinney W., Frazer D, Ma J, Schwegler-Berry D, Brain J, Castranova V., An in vivo and in vitro toxicological characterisation of realistic nanoscale CeO_2 inhalation exposures *Nanotoxicology* **in press**, DOI: 10.3109/17435390.2012.739665 (2013).

(374f) Green Pathways for Development of Nanostructured Aerogel Photocatalysts Effective At Both UV and Visible Range

Li, H., University of South Florida
Sunol, S. G., University of South Florida
Sunol, A. K., University of South Florida
Originally presented on: 11/5/2013 17:20:00 - 17:45:00

Nano-structured highly porous $TiO_2/WO_3/Fe^{3+}$ aerogel photocatalyst are prepared, characterized, and tested for model photo-catalytic reactions. The catalyst structure is tailored to capture environmental pollutants and destruct them under both u-v and visible light through oxidation.

A novel and green method is utilized to prepare the unique surfactant templated aerogel photo-catalyst that has highly porous nanostructure with high surface area and tailored pore size distribution. Sol-gel process is combined with supercritical extraction, deposition and drying. Supercritical drying with heat treatment results in titanium dioxide anatase crystal form. Templates used further enable retention and tuning of the nano-pore structure and surface properties.

The catalyst was characterized using AFM, SEM, FIB, XRD, XPS, BET and porosimetry prior to using in reactions. The band gap of the catalyst particles were determined using diffuse reflectance.

Resulting aerogel $TiO_2/WO_3/Fe^{3+}$ has similar photo-catalytic capability compared to highly optimized commercial Degussa P25 under UV exposure and offer much superior photo-catalytic capability under visible light exposure. The model reaction utilized employed Methylene Blue photo oxidation under Visible and UV light.

(375a) Dye-Sensitized Solar Cells: Using Over 100 Natural Dyes As Sensitizers

Attanayake, I., University of Moratuwa

Originally presented on: 11/5/2013 15:15:00 - 15:35:00

**DYE-SENSITIZED SOLAR CELLS:
USING OVER 100 NATURAL DYES AS SENSITIZERS**

C I.F. Attanayake [1], C. de Silva[2], B.A.J.K. Premachandra[1], A.A.P. de Alwis[1], G.K.R. Senadheera[3].*

1. *Department of Chemical & Process Engineering, University of Moratuwa, Moratuwa, Sri Lanka*

2. *Department of Chemical Engineering, City College of New York, New York, U.S.A.*

3. *Institute of Fundamental Studies, Hantana, Sri Lanka and Open University of Sri Lanka, Polgolla, Sri Lanka*

* *Corresponding Author, E-mail: ivancamilus2011@gmail.com*

ABSTRACT

Dye-sensitized Solar Cells (DSSC's) are one of the most promising new energy generation systems of photovoltaic technology to address the impending fossil fuel ? based power crisis utilizing renewable energy. These have emerged as a renewable energy source as a result of exploiting several new concepts and materials such as nanotechnology and molecular devices.

In the early 1990's Professor Michael Gratzel and his group developed a DSSC consisting of TiO_2 electrode sensitized with Ruthenium (II) Complex dye, organic liquid electrolyte with Iodine/Iodide redox couple and Platinum deposited counter electrode. In these devices a monolayer of the dye is directly attached to the semiconductor surface via carboxyl to the semiconductor surface via carboxyl group, which could realize an efficient injection of charge carriers from a photo-excited dye to the semiconductor. However, this sensitization of TiO_2 for solar applications requires not only efficient but also stable and inexpensive sensitizers. So far, several organic dyes and inorganic metal complexes have been employed to sensitize nanocrystalline TiO_2 semiconductors and one of the most efficient sensitizers is a transition metal coordination compound (Ruthenium polypiridyl complex). This complex has intense charge ? transfer (CT) absorption in the whole visible range, long excited life time and highly efficient metal-to-ligand charge transfer (MLCT).

Other organic dyes, such as phythalocyanine, cyanine dyes, xanthalene dyes, coumarin dyes etc usually perform poorly in DSSC's because of their weak binding energy with the TiO_2 film and low charge transfer absorption in the whole visible range, but these dyes are very cheap and can be prepared easily, compared to Ruthenium polypiridyl complexes. On the other hand, in nature, the fruit, flower, stem, bark, root and leaf of plants show various colours form red to purple and contain various natural dyes that can be extracted by simple procedures. Therefore, it has been emphasized by many researchers to obtain useful dyes as photosensitizers for DSSC's from natural products because of the simple preparation techniques, widely available sources and low cost. Due to these reasons, the importance of the development of low cost solar energy to electricity conversion units in principle in emphasized in this paper. Over 100 natural dyes extracted from fruits, leaves, flowers, stems, bark and roots of plants growing in Sri Lanka were used as sensitizers to fabricate dye-sensitized solar cells.

Preliminary investigation on the identification of nanocrystalline n-type TiO_2 were carried out. Fresh extracts of various fruits and vegetables were employed as sensitizers in thin layer sandwich type photo electrochemical dye-sensitized solar cells. After electrical and electronic analysis of several natural dyes of local plants, it was observed that dye extracts of Mangoostein fruit rind was found to be superior to those obtained from other dyes, and were Jsc=2.56 mA.cm^{-2}, Voc = 685.3 mV, ff = 60.02 % and h = 1.053%. Also Ekkiriya wood, Egg plant, Karawala kabilla and Banana flower yeilded Jsc, Voc, ff and h of 2.32 mA.cm^{-2}, 414.2 mV, 56.86%, 0.547% respectively ; 2.096 mA.cm^{-2}, 410.4mV, 56.42% and 0.485% respectively; 1.395mA.cm^{-2}, 443.5mV, 58.58% and 0.362% respectively; 0.763 mA.cm^{-2}, 414.0mV, 45.2% and 0.357% respectively. Whilst Fire fern leaf (not endemic to Sri Lanka) was found to display Jsc, Voc, ff and h of 4.128 mA.cm^{-2}, 405.1mV, 47.97% and 0.802% respectively.

(375b) Dye-Anchored Nanocatalysts for Improved Solar Energy Conversion Efficiency in Dye-Sensitized Solar Cells

Liu, G., University of Connecticut
Galoppini, E., Rutgers University
Agrios, A., University of Connecticut
Originally presented on: 11/5/2013 15:35:00 - 15:55:00

With the demand for sustainable and renewable energy, solar cell devices have attracted much attention which can directly convert solar energy into electrical power with the least environmental impact. Dye sensitized solar cells (DSSCs) are a very promising technology due to their attractiveness as cost-effective alternatives to conventional silicon-based photovoltaic devices. In the DSSCs, the dye is left oxidized after photo injecting an electron. A redox couple usually comprised of I^-/I_3^- in the electrolyte regenerates the oxidized dye to its neutral state and carrys the positive charge to the platinized counter-electrode. I^-/I_3^- has been commonly used as the redox couple due to the high ion mobility and slow recombination of electrons from TiO_2 to the oxidized species, triiodide (I_3^-). However, the regeneration of oxidized dye by I^- is not a simple one step reaction, which causes a large voltage loss of about 0.5–0.6 V due to the formation of diiodide radicals ($I_2^{-\bullet}$).

The talk will be focusing on reducing the overpotential of dye regeneration in DSSCs, including the preparation of Pt nanocatalyst, synthesis of new dye molecule, binding the dye to Pt nanoparticles and the assembling of TiO_2–dye–Pt system. In the earlier study, Pt nanoparticles have been prepared as the localized catalyst for the dye regeneration. A new dye molecule is synthesized with dithiolane groups for the binding to Pt nanoparticles. The aim of this talk is to demonstrate the proof-of-concept of using Pt nanoparticles to catalyze the regeneration of dye by I^- to reduce the overpotential hence improve the performance of DSSCs.

(375c) Bulk Heterojunction Polymer Solar Cells Based On Organic Semiconductor Nanomaterials

Jenekhe, S. A., University of Washington
Ren, G., University of Washington
Hwang, Y. J., University of Washington
Li, H.,
Subramaniyan, S., University of Washington
Xin, H., University of Washington
Originally presented on: 11/5/2013 15:55:00 - 16:15:00

Polymer solar cells have received increasing attention as a promising alternative for clean and low-cost solar-to-electrical energy conversion technologies. The bulk heterojunction (BHJ) photoactive layer of polymer solar cells requires rational tuning of the nanostructures towards a bicontinuous morphology to achieve maximal power conversion efficiency. The vast number of polymer semiconductors based on various donor- and acceptor-moieties has also enabled a variety of nanostructures. We have synthesized a series of new D-A copolymer semiconductors based on thiazolothiazole or benzobisthiazole building blocks and found them to be excellent p-type materials in polymer/fullerene bulk heterojunction (BHJ) solar cells with high power conversion efficiency (5-6%). Towards air-stable organic solar cells, we have also developed a series of new oligomers and polymers based on naphthalene diimide (NDI) moieties that function as efficient acceptor materials in BHJ polymer solar cells. BHJ solar cells based using NDI acceptors have showed one of the highest photovoltaic efficiencies achieved to date in *non-fullerene* polymer solar cells. Efficient all-nanowire BHJ solar cells were fabricated based on donor polymer nanowires and acceptor oligomer nanowires. Our results show that donor-acceptor architecture is a powerful means of tuning the molecular structures and nanostructures of polymer/oligomer semiconductors and their photovoltaic properties toward achievement of highly efficient polymer solar cells.

(375d) Engineering Energy Levels At the TiO2:P3HT Interface Using Atomic Layer Deposition

Dorman, J., University of California at Los Angeles
Weickert, J.,
Putnik, M.,
Schmidt-Mende, L.,
Originally presented on: 11/5/2013 16:15:00 - 16:35:00

It has been shown over the last two decades that conversion efficiencies up to 12% can be achieved using TiO_2 based solar cells, such as in liquid electrolyte dye sensitized solar cell. Recently, there has been a push to replace these hazardous electrolytes with organic materials to create environmentally friendly devices with extended lifetimes. However, one of the limitation of these hybrid solar cells is electron-hole interaction across the metal-oxide and organic hole transporter interface.

In this work, a core-shell nanostructured hybrid solar cells is synthesized with a single crystal core in order to increase electron mobility and light scattering without additional recombination effects. A combination of hydrothermal and atomic layer deposition (ALD) was used to fabricate TiO_2 nanowires (NWs) in order to take advantage of the directed electric field within the structures. During the growth process a Sn^{4+} dopant is introduced to create a doped $Sn:TiO_2$ nanostructured array with increased electron mobility. After the NW growth, an additional ALD step is used to deposit a TiO_2 layer, creating the core-shell structure. The dopant gradient within the core-shell structure causes the electrons to migrate toward the core of the nanowire due to the lower energy conduction band, potentially reducing electron-hole recombination at the TiO_2:P3HT interface. The conduction band engineering has a similar effect as that seen with dipole modification of the interface. Additionally, the surface can be further modified with dye molecules. This doped core-shell structure has resulted in a conversion efficiency of 2 % with a surface treatment of the squaraine dye SQ2. This efficiency is due to the contribution of the dye and the P3HT in the photon current generation.. The engineered energy levels and interfacial modifiers have a significant effect on the external quantum efficiency and internal resistances, as determined using various characterization methods.

(375e) Layered Double Hydroxides As An Effective Additive in Polymer Gelled Electrolyte Based Dye-Sensitized Solar Cells

Ho, H. W., National Tsing Hua University
Wei, T. C., National Tsing Hua University
Lu, S. Y., National Tsing-Hua University
Originally presented on: 11/5/2013 16:35:00 - 16:55:00

Dye-sensitized solar cells (DSSC) continue to draw a great deal of research attention as a promising alternative clean energy device, because of the simplicity and low cost involved in the manufacturing process and the reasonable power conversion efficiency (PCE). The boost in PCE to 12.3 % reported by Yella et al. [1] in 2012 makes DSSCs even more competitive. One of the key issues yet to be fully resolved is the volatility, leakage, and poor long term stability associated with the use of liquid electrolytes in DSSCs. Intensive and extensive research efforts have been devoted to the development of gel or solid electrolytes to tackle the issue. For solid electrolytes, the poor contact at the electrolyte/electrode interface often hampers the boost in PCE. Gel electrolytes, on the other hand, possess the advantages of low volatility and superior long term stability as compared with liquid electrolytes, and of good electrolyte/electrode contacts as compared with solid electrolytes. In the past few years, use of polymer gelled electrolytes containing performance boosting inorganic additives has been a promising approach. Particularly, clay materials, because of their high chemical stability, ion exchange ability, and natural abundance, have been a popular candidate as the inorganic additive [2]. Ho and co-workers added exfoliated montmorillonites to polymer gelled electrolytes to reduce the electrolyte/electrode interfacial impedances and to enhance the reduction rate of I_3^- at the cathode [3,4]. Geng et al. found that addition of montmorillonites to poly(ethylene oxide) (PEO) gelled electrolytes can inhibit the charge recombination at the anode [5]. Lai et al. used polyvonydiene fluoride-cohexafluoro propylene (PVDF-HFP) gelled electrolytes with mica nanoparticles as the additive to reduce the crystallinity of the polymer, leading to higher PCEs [6].

The above mentioned layer-structured materials, montmorillonites and mica, belong to the class of cationic clays, carrying negative charges on their hydroxide layers counter-balanced by positive inter-layer ions. The popular electrolyte redox couple for DSSCs, I^-/I_3^-, however, contains anions. The use of anionic clay materials, containing inter-layer cations, may prove advantageous for DSSCs [2]. In this work, layered double hydroxides (LDH), a class of anionic clay materials, were developed as the inorganic additive for PVDF-HFP gelled electrolytes. ZnAl-CO_3-LDH and ZnAl-Cl-LDH were prepared with a separated nucleation and aging method and low supersaturation co-precipitation method, respectively. The addition of the two LDHs, in terms of PCEs, significantly improved over the plain PVDF-HFP gelled electrolyte and competed favorably with the liquid electrolyte, 8.13% for the liquid electrolyte, 7.48% for the PVDF-HFP gelled electrolyte, 8.11% for the ZnAl-CO_3-LDH/PVDF-HFP gelled electrolyte, and 8.23% for the ZnAl-Cl-LDH/PVDF-HFP gelled electrolyte based DSSCs. The good performance in PCE achieved by the LDH based DSSCs was mainly attributed to the significant boost in open circuit voltages (Voc), from 0.74 V for both the liquid electrolyte and PVDF-HFP gelled electrolyte based DSSCs to 0.79V for the ZnAl-CO_3-LDH/PVDF-HFP gelled electrolyte based DSSC and 0.84V for the ZnAl-Cl-LDH/PVDF-HFP gelled electrolyte based DSSC. The boost in Voc was found to come from the positive shift in the redox potential of the electrolyte redox couple as revealed from a cyclic voltammetry analysis. As for the long term stability, a PCE retention rate of 89% after 624 h was achieved with the ZnAl-CO_3-LDH/PVDF-HFP gelled electrolyte based DSSC, much better than that of the liquid electrolyte based one. References:

1. A. Yella, H.-W. Lee, H.-N. Tsao, C. Yi, A.-K. Chandiran, M.-K. Nazeeruddin, Eric W.-G. Diau, C.-Y. Yeh, S.-M. Zakeeruddin, M. Grätzel, Science 334, (2011)
2. X.-Wang, S.-A. Kulkarni, B.-I. Ito, S.-K. Batabyal, K. Nonomura, C.-C. Wong, M. Gratzel, S.-G. Mhaisalkar, S. Uchida, ACS Appl. Mater. Interfaces, 5(2013)
3. C.-W. Tu, K.-Y. Liu, A.-T. Chien, M.-H. Yen, T.- H. Weng, K.-C. Ho, K.-F. Lin , Journal of Polymer Science: Part A: Polymer Chemistry, 46(2008)
4. C.-H. Lee , K.-Y. Liu, S.-H. Chang, K.-J. Lin, J.-J. Lin, K.-C. H, K.-Fu Lin, Journal of Colloid and Interface Science ,363 (2011)
5. Y. Geng, Y. Shi, L. Wang, B. Ma, R. Gao, Y. Zhu, H. Dong and Y. Qiu, Phys. Chem. Chem. Phys.,13 (2011)
6. Y.-H. Lai, C.-Y. Lin, J.-G. Chen, C.-C. Wang, K.-C. Huang, K.-Y. Liu, K.-F. Lin, J.-J. Lin, K.-C. Ho, Solar Energy Materials & Solar Cells ,94 (2010)

(375f) Boron-Modified TiO2 Microsphere With Enhanced Photovoltaic Performance

Rao, J., Nanjing University of Technology
Zhou, X., Nanjing University of Technology
Originally presented on: 11/5/2013 16:55:00 - 17:15:00

Nanotubes-assembled rutile TiO_2 microspheres were successful prepared by a facile two-step hydrothermal approach in the presence of sodium metaborate tetrahydrate. SEM and FESEM micrographs revaled that the addition of boron enrichs the assembled nanorodes with a reduced diameter during hydrothermal process, probably result from the increased crystal defects induced more reaction sites for crystal growth after introduction of lots of boron compound. A slight shift to the left and weakened intensity of the diffraction peaks appeared in XRD pattern without obvious peaks for diboron trioxide evidenced the obtained rutile TiO_2 accompanied with boron at interstitial sites. When applied to the photoanode of dye-sensitized solar cell, the photovoltaic device achieved a favorable photoelectricity conversion efficiency of 3%, which was 74% higher than that of the non-modified TiO_2. Further investigation on the cells' electrochemical properties were adopted by the impedance spectroscopy, which showed a longer electron lifetime and lower electron transfer rate at photoanode/dye/electrolyte interface compared to the pure TiO_2 sample.

(375g) Effect of Ionic Liquid Electrolytes in DSSCs With Titanium Dioxide (TiO2) Inverse Opal Structures

Ramesar, N., CUNY City College of New York
Kretzschmar, I., The City College of New York, The City University of New York
Originally presented on: 11/5/2013 17:15:00 - 17:35:00

Effect of Ionic Liquid Electrolytes in DSSCs with Titanium Dioxide (TiO_2) Inverse Opal Structures

Naomi Ramesar and Ilona Kretzschmar

Department of Chemical Engineering, CUNY City College, 160 Convent Avenue, New York, NY 10031

AIChE Conference 2013: ***Abstract***

Dye-sensitzed solar cells (DSSC) are low-cost alternatives to conventional solar cells that can work well in low light conditions. Despite considerable study on improving the efficiency of DSSCs, the current liquid electrolyte cell has plateaud at 11.1%. A major problem with these cells is the low viscosity and high volatility of the electrolyte, i.e., acetonitrile, which cause leakage and volatilization. We propose that using ionic liquids (ILs), which are more viscous, less volatile, and conductive, may be more suitable electrolytes. However, the higher viscosity of ILs may cause incomplete infiltration of the DSSC's nanoporous TiO_2 electrode. Here, we present a study of DSSCs with TiO_2 inverse opal electrodes of controlled pore sizes (0.1-1 mm) and ionic liquid derivatives of 1-alkyl-3-methylimidazolium tetrafluoroborate (alkyl: ethyl, butyl, and decyl) with viscosities ranging from 25.2 to 721 cP. The stability and functionality of the DSSC is tested using current-voltage characteristics.

(375h) Porous Nanowire Assemblies On Graphene for Dye-Sensitized Solar Cells

Charpentier, P. A., Western University
Alsharari, Q., Western University
Farhangi, N., Western University
Moula, G., University of Western Ontario
Ayissi, S., Western University
Originally presented on: 11/5/2013 17:35:00 - 17:55:00

Tremendous interest exists towards synthesizing nanoassemblies for dye-sensitized solar cells (DSSCs) using earth abundant and friendly materials with green scalable approaches.[1] Traditional approaches towards nanowires prevent multi-metallic nanowires, give low surface areas, and give a lack of surface chemistry for decorating nanoparticles. Supercritical CO_2 ($scCO_2$) is an enabling solvent to overcome several of these challenges for the synthesis of nanomaterials and nanodevices due to its low viscosity, high diffusivity, and "zero" surface tension [2]. This work focused on synthesizing titanium dioxide porous and metal-doped titanium dioxide nanowires from graphene surfaces using $scCO_2$. [3,4]. Graphene was examined as a catalyst mat for DSSC assemblies while quantum dots were decorated on the TiO_2 nanowires.

The results showed that undoped and doped (Fe, Zr, various quantum dots) TiO_2 nanowires were grown on the surface of graphene sheets using a sol-gel method in $scCO_2$, which helped exfoliate graphene sheets. TiO_2 nanowires less than 40 nm (when $scCO_2$ was used as a solvent) were uniformly decorated on the graphene sheets, in which surface -COOH functionalities acted as a template for nanowire growth. A variety of bi and trimetallic nanowires were synthesized, giving enhanced optical properties and photoefficiencies in DSSCs. Fe and Zr doped TiO_2 nanowires grown on graphene showed smaller crystal sizes, gave higher visible absorption and surface areas, and higher efficiency solar cells (>4% efficiency).[5] HRTEM images confirmed formation of porous nanowires with high aspect ratios and high metal dispersion using XPS. EELS mapping showed high dispersion of bimetallic and multi-metallic wires that could be connected directly to graphene assemblies. Higher surface areas of bimetallic and multi-metallic nanowires on graphene were found to enhance the electron-hole separation while reducing the band gap of the nanowires making them a potential for DSSCs. [3]. Theoretical band structures of assemblies were studied using the Vienna ab-initio Simulation Package (VASP) based on the Density Functional Theory (DFT). Functionalized graphene sheets decreased significantly the band gap but increase the binding energy values due to carboxylate adsorption sites, hence enhancing DSSC performance.[5]

1) Oudel, P., & Qiao, Q. (2012). One dimensional nanostructure/nanoparticle composites as photoanodes for dye-sensitized solar cells. Nanoscale, 4(9), 2826-2838.

2) R. Sui, Paul A. Charpentier*, Synthesis of Metal and Silicon Oxides Nanostructures by Direct Sol-Gel Reactions in Supercritical Fluids, *Chem. Rev.*, (2012), 112(6), 3057-3082..

3) Nasrin Farhangi, Yaocihuatl Medina-Gonzalez , Rajib Roy Chowdhury, and Paul A. Charpentier, Using Supercritical CO_2 for growing TiO_2/Graphene Nanoassembles: Synthesis and Photocatalytic Effect, *Nanotechnology* (2012), 23(29), 294005/1-294005/12.

4) Nasrin Farhangi, Rajib Roy Chowdhury, Yaocihuatl Medina-Gonzalez, Madhumita B. Ray and Paul A. Charpentier Visible light active Fe doped TiO2 nanowires grown on Graphene using Supercritical CO_2, Applied Catalysis B: Environmental 110 (2011) 25– 32.

5) Ayissi, Serge; Charpentier, Paul; Farhangi, Nasrin; Wood, Jeffery; Palotás, Krisztián; Hofer, WernerInteraction of Titanium Oxide Nanostructures with Functionalized Graphene Sheets: A DFT Study, J. Phys Chem. C (2013) Submitted.

(436a) Hydrogen Production From Water

McDaniel, M. D.,
Ekerdt, J. G., University of Texas at Austin
Originally presented on: 11/6/2013 8:30:00 - 9:00:00

Silicon-based photocathodes show promising potential for photoelectrochemical (PEC) water splitting for renewable production of hydrogen. Trade-offs between stability and efficiency have limited the success of Si-based photoelectrodes. To overcome this issue, we utilize a thin layer (≈ 2 nm) of strontium titanate, $SrTiO_3$ (STO), which is grown epitaxially on p-doped Si (001) by molecular beam epitaxy. The STO layer greatly enhances the stability of the Si-based photocathode. In addition, the high quality STO layer and clean, low defect interface between STO(001) and Si(001) provides minimal trap states for charge carriers. Moreover, there is no conduction band offset between STO and Si. This allows the photo-generated electrons to easily move from the Si substrate to the STO surface, where hydrogen ions are reduced to form gaseous hydrogen. Based on our electrochemical measurements, we find that the STO-Si photocathode is both stable and exhibits high conversion efficiency. Studies with continuous Ti/Pt films for H_2 evolution yield a current density above 15 mA/cm^2; patterned metal layers are expected to give higher current densities. The results indicate that Si-based photocathodes for PEC water splitting can be improved using a thin STO layer, providing a viable method for clean hydrogen production.

(436b) Core-Shell Photoelectrochemical Electrodes for Water Splitting

Peng, Q., Duke University
Kalanyan, B., North Carolina State University
Hoertz, P.,
Miller, A.,
Kim, D. H., North Carolina State University
Hanson, K., UNC-Chapel Hill
Alibabaei, L., UNC-Chapel Hill
Liu, J., Duke University
Meyer, T., UNC-Chapel Hill
Parsons, G. N., North Carolina State University
Glass, J. T., Duke University
Originally presented on: 11/6/2013 9:00:00 - 9:18:00

Photoelectrochemical (PEC) processes offer one of most promising ways to convert solar energy into fuels (e.g. H_2, and CH_3OH), especially H_2. In order to synthesize high efficient and stable PEC water splitting electrodes, one must optimize the light absorption, charge separation, and charge transfer processes of the electrode simultaneously. Core-shell nanostructures with transparent conductive cores and semiconductor shells show great promise for optimizing PEC performance of photoactive materials, such as TiO_2. However, limited research has been performed in this field.

We will present the result of using transparent conductive network consisting of low cost, solution-processed antimony doped tin oxide nanoparticles (nanoATO) and tin doped indium oxide nanoparticles (nanoITO) for enabling core-shell TiO_2 PEC electrodes. In detail, we will present the synthesis of TiO_2 PEC electrodes by coating solution-processed nanoATO on FTO glass with TiO_2 through atomic layer deposition (ALD). The conductive, porous nanoATO film-supported TiO_2 electrodes, yielded a highest photocurrent density of 0.58 mA/cm^2 under AM 1.5G simulated sunlight of 100 mW/cm^2. This is approximately 3× the maximum photocurrent density of planar TiO_2 PEC electrodes on FTO glass. Transient photocurrent measurements showed that nanoATO films reduce charge recombination by accelerating transport of photoelectrons through the less defined conductive porous nanoATO network. Further, we will show that the effect of interfacial reaction during the TiO_2 ($TiCl_4/H_2O$) ALD on transparent nanoATO and nanoITO on their PEC performance. The scalable solution-processed porous transparent nanoATO and nanoITO films are promising as a framework to replace other conductive scaffolds (Si nanowire forest, CNTs, and $TiSi_2$ nanonets) for PEC electrodes. The results are not only valuable for PEC, but also informative for the related DSSC research.

(436c) Design of Photocatalysts Active in Infrared and Visible Range

Lee, D. C., Korea Advanced Institute of Science and Technology (KAIST)
Originally presented on: 11/6/2013 9:18:00 - 9:36:00

Infrared- and visible-active semiconductors are promising candidates for light-harvesting materials. When the semiconductor nanocrystals are within quantum confinement regime, they exhibit size tunable bandgap, enabling light-harvesting even more controllable. Furthermore, the dynamics of excited states can also be controlled through a design of heterostructure semiconductor nanocrystals. We synthesized PbSe/CdSe/CdS core/shell/shell nanocrystals with varying structural design, e.g., spheres, pyramid and tetrapod. These nanocrystals show varying electron-hole exchange interactions and radiative recombination lifetimes depending on their dimensions. We also prepared semiconductor/metal nanojunction and used the nanomaterials as photocatalysts for reduction of various dyes by growing metal tips at the corners of semiconductor nanocrystals. Interestingly, the morphology change leads to the difference in the tip growth pattern. In addition, the growth of metal particles affects the Fermi level of the entire hybrid composites, leading to the change in their photocatalytic activity. In the presentation, the implications of the heterostructure quantum dots and their hybrid composites with metal nanoparticles will be discussed in the context of energy applications.

(436d) Efficient Photoelectrochemical Water Splitting With Si-Based Metal-Insulator-Semiconductor Photoelectrodes

Esposito, D. V., University of Delaware
Lee, Y., National Instittue of Standards and Technology
Talin, A. A., Sandia National Laboratory
Moffat, T. P., National Institute of Standards and Technology
Originally presented on: 11/6/2013 9:36:00 - 9:54:00

Photoelectrochemical (PEC) water splitting is an attractive pathway for renewable production of hydrogen, but the efficiency and stability of semiconducting photoelectrodes must be improved. One promising approach to achieve both high efficiency and good electrochemical stability is the metal-insulator-semiconductor (MIS) photoelectrode architecture.[1,2] The MIS design consists of catalytic metal structures, or collectors, deposited on a 1-3 nm thick oxide-covered semiconductor. Of great importance to this design is the thin insulating oxide layer, which must protect the semiconductor from the potentially corrosive electrolyte while simultaneously mediating minority carrier tunneling between the semiconductor and collectors. In recently published work,[2] we have demonstrated significant improvement in the performance of Si-based MIS photocathodes through interfacial engineering. These planar photocathodes exhibit excellent stability and appreciable photoelectrode conversion efficiencies, up to 2.9%, but further improvement in efficiency must be achieved in order to make this process economically viable.

In this work, nano- and micro-structured Si-based MIS photoelectrodes fabricated by reactive ion etching and vapor-liquid-solid chemical vapor deposition are explored as a means to further increase the efficiency of Si-based photoelectrodes. The use of 3D-structured electrodes offer several potential advantages for achieving higher efficiency, including increased surface area for catalysis, improved collection of photo-generated carriers, and decreased optical reflection losses. However, the use of 3D photoelectrodes also introduces detrimental effects that create trade-offs between optical, catalytic, and electronic properties. In this talk, design principles for optimizing the performance of 3D-structured photoelectrodes are described and a comparison of the performance of planar and 3D-structured MIS photoelectrodes is presented. The incorporation of MIS photoelectrodes into stand-alone PEC water splitting devices will also be discussed. Calculations based on the method of detailed balance show that PEC devices utilizing Si-based MIS photoelectrodes are capable of achieving solar-to-hydrogen conversion efficiencies greater than the 2015 efficiency target of 15% set by the Department of Energy.[3]

References

[1.] H.J. Lewerenz, et al., *Electrochem. Acta*, **2011**, 56, 10726.

[2.] D.V. Esposito, I. Levin, T.P. Moffat, A.A. Talin, *Nature Materials*, **2013**, (in press).

[3.] Fuel Cells Technologies Office Multi-Year Research, Development and Demonstration Plan. US Department of Energy Efficiency and Renewable Energy. Available from: http://www1.eere.energy.gov/hydrogenandfuelcells/mypp/ (2012)

(436e) Graphene Oxide Photocatalysts for Water Splitting and Its Upconverted Photoluminescence

Yeh, T. F.,
Teng, H.,

Originally presented on: 11/6/2013 9:54:00 - 10:12:00

Graphene-Oxide Photocatalysts for Water Splitting and Its Upconverted Photoluminescence

Te-Fu Yeh[1] and Hsisheng Teng[1,2,*],

[1]*Department of Chemical Engineering and Research Center for Energy Technology and Strategy, National Cheng Kung University, Tainan 70101, Taiwan*

[2]*Center for Micro/Nano Science and Technology, National Cheng Kung University, Tainan 70101, Taiwan*

**E-mail:* *hteng@mail.ncku.edu.tw*

Abstract

Graphene oxide (GO) is a p-type semiconductor with a suitable band positions for photocatalytic watersplitting. GO sheets have a hydrophilic 2D carbon structure that allows for extensive chemical modification. This study presents strategies of tuning the conductivity type of GO from p-type to n-type for simultaneous evolutions of H_2 and O_2 under visible light irradiation. Further converting the n-type graphene oxide sheets to quantum dots results in stoichiometric evolution of H_2 and O_2 at 2-to-1 ratio. The quantum dots also serve as an energy-converting medium with upconverted photoluminescence to promote the activity of metal-oxide photocatalyts.

Introduction

Polymeric semiconducting molecules, which have a large contact area with water, attract significant attention as alternatives to metal-containing photocatalysts. Graphene oxide (GO) derived from graphite oxidation is a molecule-like semiconductor [1]. GO can be extensively dispersed in water to molecular scale and has a large exposed area. Graphene is a zero gap semiconductor whereas the GO gap increases with the oxidation level [2,3]. The bonding with electron-withdrawing oxygen functionalities results in the p-type conductivity of GO. N-type conductivity appears when graphene covalently bonds to electron-donating nitrogen functionalities [4-6].

In addition to chemical doping, we exploit the quantum-confinement effect to modulate electronic properties. Unlike two-dimensional graphene, the zero-dimensional graphene quantum dots (GQDs) have a finite band gap depending on their size [7]. The confined p-electron in sp^2 domain gives the quantized discrete levels [8]. With these features, GQDs exhibit upconvered photoluminescence (PL) [9].

This study presents the potential of GO as a photocatalyst for water splitting. P-type GO can catalyze H_2 evolution with the presence of electron donors. Ammonia-treated GO (NGO) exhibits n-type conductivity and promising photocatalytic activity for generating H_2 and O_2 simultaneously, but with the ratio below the stoichiometric value (that is, 2). Miniaturized N-doped graphene oxide quantum dots (NGO QDs) function as photocatalysts for stoichiometric H_2 and O_2 generation from water cleavage under visible light irradiation. We also demonstrate the upconversion ability of NGO QDs.

Experimental

GO was prepared using a natural graphite powder through a modified Hummers' method. NGO was obtained through nitridation of the as-prepared GO performed at room temperature using a flow of NH_3 gas. NGO QDs were prepared from oxidation of nitrogen-doped graphene. Photocatalytic reactions were conducted at approximately 25 Â¢XC in a gas-enclosed inner irradiation system or a side irradiation system.

Result and Discussion

We demonstrated the photocatalytic activity of GO in the gas-enclosed circulation system with mercury-lamp irradiation. Continuous H_2 evolution over metal-free GO was observed for reactions conducted in pure water and an aqueous methanol solution [1]. The total evolution of H_2 from aqueous methanol solution far exceeds the amount of H_2 obtained from pure water. GO cannot catalyze O_2 generation even with the presence of the electron scavenger Ag^+ ion.

We modified the as-synthesized GO photocatalyst by ammonia treatment to form NGO, which exhibited n-type conductivity. The change of conductivity type was contributed by the electron-donating property of the introduced amino groups. NGO effectively catalyzes O_2 evolution under mercury-lamp irradiation [6]. The n-type conductivity of NGO attracts hole migration toward the interface region for O_2 evolution. NGO gave rise to simultaneous evolutions of H_2 and O_2 from the side irradiated system under visible light irradiation [6]. The molar ratio of the evolved H_2 and O_2 was smaller than the stoichiometric ratio 2-to-1.

Quantum confinement effect elevates the conduction band edge, promoting the electron injection for H_2 evolution from water. We found stoichiometric evolutions of H_2 and O_2 from pure water over NGO QDs with visible-light irradiation. Figure 1 shows the PL spectra of NGO QDs excited by long-wavelength light (660-900 nm), with the upconverted emission located in the range of 450-530 nm. The upconverted PL emission of NGO QDs excited $NaTaO_3$ to catalyze H_2 evolution from an aqueous methanol solution under visible light irradiation.

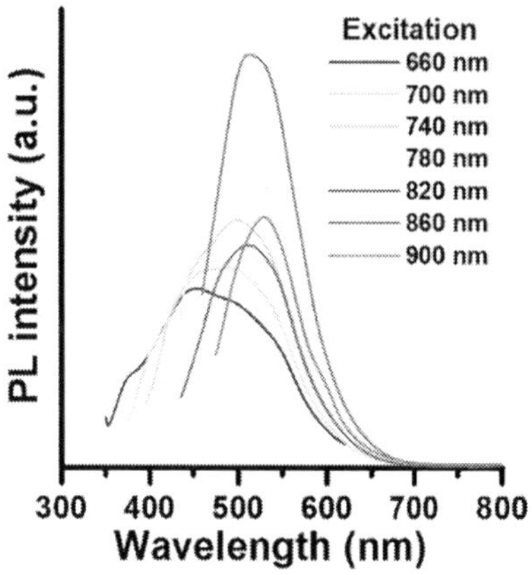

Figure 1. Upconverted PL from NGO QDs. The excitation wavelength ranges from 660 nm to 900 nm. The upconverted emission located in the range of 450-530 nm.

Conclusions

The p-type GO catalyzed H_2 evolution from an aqueous methanol solution under mercury-lamp illumination. The introduction of nitrogen functionalities converted the p-type GO to an n-type NGO, which was capable of catalyzing simultaneous H_2 and O_2 evolutions from water under visible light. The NGO QD catalyst achieved overall water splitting ($H_2/O_2 = 2$) under visible light irradiation. The upconversion ability of NGO QDs made wide-gap photocatalysts active under visible-light illumination.

Acknowledgements

This research is supported by the National Science Council of Taiwan (101-2221-E-006-243-MY3 and 101-2221-E-006-225-MY3) and the "Aim for the Top-Tier University and Elite Research Center Development Plan" of National Cheng Kung University.

References

[1] T.F. Yeh, J.M. Syu, C. Cheng, T.H. Chang, H. Teng, *Adv. Funct. Mater.* **2010**, 20, 2255.

[2] T.F. Yeh, F.F. Chan, C.T. Hsieh, H. Teng, *J. Phys. Chem. C* **2011**, 115, 22587.

[3] T.F. Yeh, J. Cihlář, C.Y. Chang, C. Cheng, H. Teng, *Materials Today* **2013**, 16, 78.

[4] X. Wang, X. Li, L. Zhang, Y. Yoon, P. K. Weber, H. Wang, J. Guo, H. Dai, *Science* **2009**, 324, 768.

[5] C. C. Hu, H. Teng, *J. Phys. Chem. C* **2010**, 114, 20100.

[6] T.F. Yeh, S.J. Chen, C.S. Yeh, H. Teng, *J. Phys. Chem. C* **2013**, 117, 6516.

[7] L. A. Ponomarenko, F. Schedin, M. I. Katsnelson, R. Yang, E. W. Hill, K. S. Novoselov, A. K. Geim, *Science* **2008**, 320, 356.

[8] G. Eda, Y. Y. Lin, C. Mattevi, H. Yamaguchi, H. A. Chen, I. S. Chen, C. W. Chen, M. Chhowalla, *Adv. Mater.* **2010**, 22, 505.

[9] H. Li, X. He, Z. Kang, H. Huang, Y. Liu, J. Liu, S. Lian, C. H. A. Tsang, X. Yang, S. T. Lee, *Angew. Chem. Int. Ed.* **2010**, 49, 4430.

(436f) The Study of Cu2ZnSnS4 Nanocrystal/TiO2 Nanorod Arrays Heterojuction Photoelectrochemical Cell for Hydrogen Generation

Ho, T. Y.,
Chen, L. Y.,
Originally presented on: 11/6/2013 10:12:00 - 10:30:00

In this study, we reported TiO_2 nanorods arrays (TiO_2 NRAs) synthesized by hydrothermal method with controllable densities, diameters and lengths for photoelectrochemical hydrogen generation. The densities and diameters of the TiO_2 NRAs could be easily controlled by the concentration of titanium precursor (titanium isopropoxide, TTIP) and the amount of the HCl. With different reaction time, the lengths of TiO_2 NRAs could be increased from 2 μm to 9 μm. The photocurrent density and conversion efficient were increased with the lengths of TiO_2 NRAs. The maxima current density and conversion efficiency are 2.5mA/cm^2 at 1.23V vs RNHE and 0.95% with 4-5 μm in length under 100mA/cm^2 Xenon lamp illumination with AM1.5 filter, respectively. To improve the conversion efficiencies, various sensitizers, such as dye or quantum dot, are usually used to cover the absorption in the visible light range. Herein, Cu_2ZnSnS_4 (CZTS), which owns direct band gap semiconductor with 1.5eV and suitable band positions, was employed to extend the absorption range. In this study, CZTS nanoparticles were direct decorated on TiO_2 NRAs by solvothermal method with metal chloride and thiourea as precursor and ethanol as solvent to improve the conversion efficiency for hydrogen generation.

(436g) Development of New Generation of Sub-Nanometer Catalysts for Sustainable Energy Applications

Wu, Q., Stony Brook University/BNL
Shen, P., Stony Brook University / BNL
Zhao, S., Stony Brook University
Ramakrishnan, G.,
Su, D., Brookhaven National Laboratory
Li, Y.,
Orlov, A., Stony Brook University / BNL
Originally presented on: 11/6/2013 10:30:00 - 10:48:00

Developing novel heterostructured nanomaterials for energy applications as water splitting under visible light has enormous potential for energy production. Currently the existing materials have very low quantum efficiency, which indicate that new approaches are required in catalysts design. In this work we have developed and characterized sub nanometer Au and Pt nanoparticles, which greatly enhanced rate of hydrogen production of CdS. These heterostructured materials were characterized by TEM, STEM, XRD, UV-Vis spectroscopy and DRIFTS. We have also conducted comprehensive studies of these catalysts in photocatalytic reactors, where we observed a substantial hydrogen production under visible light.

(436h) ZrO2 Incorporated ZnO/TiO2 Mesoporous Photocatalyst for H2 Generation

Hussein, A., South Dakota Shool of Mines and Technology
Shende, R., South Dakota School of Mines and Technology
Originally presented on: 11/6/2013 10:48:00 - 11:06:00

The present research deals with the synthesis of ZrO_2 incorporated coupled ZnO/TiO_2 photocatalyst and its effectiveness on H_2 generation from aqueous methanol solution. Ordered mesoporous ZrO_2 supported coupled TiO_2/ZnO materials were synthesized using surfactant as structure directing agent, titanium isopropoxide, zinc acetate dihydrate and zirconium n-propoxide precursors. The ZrO_2 incorporated coupled ZnO/TiO_2composites were characterized by XRD, UV-Vis spectra, SEM/TEM, XRF, Raman and N_2 adsorption-desorption isotherm. Photocatalytic hydrogen evolution was carried out in a 7 mL quartz reactor at ambient temperature. The focused intensity and area on the quartz cuvette were 0.60 W and 2 cm^2, respectively. In a typical experiment, 2.0 mg of ZrO_2/ZnO-TiO_2 photocatalyst was dispersed in 3.5 mL of H_2O containing 0.5 mL methanol or 10 mg biomass as a sacrificial agent. A 300 W xenon lamp (Oriel) equipped with a UV-filter (240 nm - 400 nm) was employed as the light source. Prior to irradiation, the suspension of the catalyst was degassed with ultrapure nitrogen for 30 min to completely remove the dissolved oxygen and to ensure the reactor in an anaerobic condition. From 3 mL reactor overhead space, 0.1 mL of gaseous samples were withdrawn periodically and analyzed using the gas chromatograph (GC), from which photocatalytic effectiveness towards H_2 generation was evaluated and compared with coupled TiO_2/ZnO material.

(499a) SiC Photoanodes for Solar Water Splitting

Bohn, C., National Institute of Standards and Technology
Sharifi, F.,
Kang, M. G.,
Szalai, V., National Institute of Standards and Technology
Originally presented on: 11/6/2013 12:30:00 - 12:49:00

AIChE 2013 Session: Nanomaterials for Hydrogen Production and Fuel Cells

Title: SiC Photoanodes for Solar Water Splitting

Authors: C.D. Bohn, M.G. Kang, F. Sharifi, V.A. Szalai

SiC ceramics offer a rich field for exploration for solar water splitting.Â Polytypes include 4H-SiC with a bandgap of 3.23 eV, 6H-SiC with a bandgap of 3.05 eV and 3C-SiC with a bandgap of 2.36 eV.Â Single crystal substrates of some of these materials were obtained from commercial vendors.Â The effect of nanoscale morphology and the addition of Pt nanocrystals at the interface between the SiC and electrolyte were systematically investigated in a pH=7 potassium phosphate buffer under AM 1.5 illumination.Â Gas chromatography provided insight into the chemical products formed by the n-type SiC photoanodes.Â Additionally, linear sweep voltammetry, impedance spectroscopy and scanning electron microscopy were used to monitor the structural evolution of the semiconductor surface.Â Results are discussed in the broader context of photoelectrochemical hydrogen production and recommendations for improving the efficiency of water splitting with SiC are proposed.

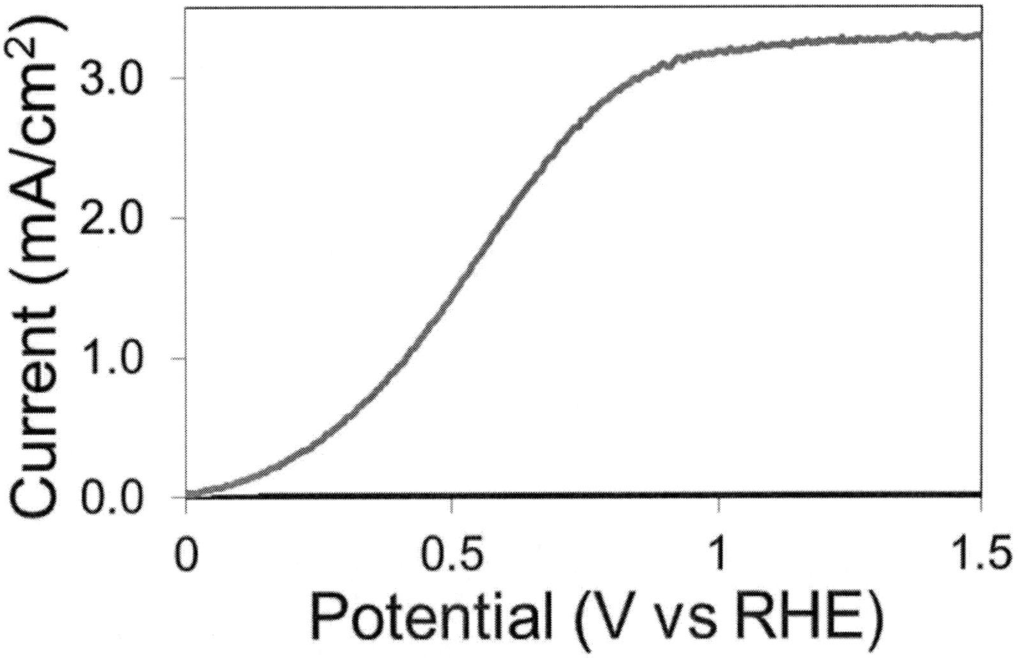

Keywords: solar water splitting, SiC, photoanode

(499b) Advanced Electrocatalysts From Organic Solution Synthesis

Wang, C., Johns Hopkins University

Originally presented on: 11/6/2013 12:49:00 - 13:08:00

Alloy nanoparticles (NPs) have attracted increasing interest due to their superior performance in various catalytic applications. For example, Pt alloys with transition metals (PtM with M = Fe, Co, Ni, etc.) have been found to be highly active for the oxygen reduction reaction in fuel cells. A lot of efforts have been dedicated to the synthesis of alloy catalysts, which are usually in the form of alloy NPs dispersed in high surface area carbon matrix. The approaches mostly include co-precipitation of metal salts in aqueous solution, impregnation, and electrodeposition. In spite of the successful preparation of various types of alloy catalysts, it yet remains challenging to synthesize monodisperse and homogeneous alloy catalysts with robust control over size and composition. Here we introduce our organic solvothermal approach towards high-quality alloy nanomaterials for electrocatalytic applications. Control over particle size, alloy composition and alloy homogeneity has been achieved by matching the different reduction and growth rates of the alloying elements, which has enabled systematic studies of the effects of these parameters on electrocatalysis. Our studies thus provide clues for the design and optimization of alloy nanomaterials for catalytic applications.

(499c) Nanostructured Ti -Fe2O3/Cu2O Bilayered Thin Films for Photoelectrochemical Water Splitting

Sharma, D., Dayalbagh Educational Institute,Dayalbagh
Upadhyay, S., Dayalbagh Educational Institute
Choudhary, S., Dayalbagh Educational Institute
Satsangi, V. R., Dayalbagh Educational Institute
Shrivastav, R., Dayalbagh Educational Institute
Dass, S., Dayalbagh Educational Institute
Originally presented on: 11/6/2013 13:08:00 - 13:27:00

Nanostructured Ti-Fe_2O_3/Cu_2O Bilayered Thin Films for Photoelectrochemical Splitting of Water

Dipika Sharma, Sumant Upadhyay, Surbhi Choudhary, Vibha R. Satsangi?,

Rohit Shrivastav and Sahab Dass,

Department of Chemistry, ?Department of Physics & Computer Sciences

Â Faculty of Science, Dayalbagh Educational Institute (Deemed University),

Agra-282110 (India)

Nanostructured thin films of pure Fe_2O_3, Ti ?Fe_2O_3, Cu_2O, and bilayered Fe_2O_3/Cu_2O, Ti-Fe_2O_3/Cu_2O were deposited on ITO glass substrate using spray pyrolysis method.Â Among various metal oxide semiconductors, cuprous oxide Cu_2O [Eg=2.4eV] and Iron oxide Fe_2O_3 [Eg=2.1eV] were chosen for the present study on account of being abundant, non-toxic, environmentally safe, having low band gap and possess staggered type band edges which is suitable for efficient charge separation in water splitting process.

All samples were characterized using XRD, SEM and UV?Vis spectrometry. Photo electrochemical properties were also investigated in a three-electrode cell system. UV-Vis absorption spectrum for pure Fe_2O_3, Ti ?Fe_2O_3, Cu_2O, and Fe_2O_3/Cu_2O, Ti-Fe_2O_3/Cu_2O bilayered thin films shows absorption in visible light region. Nanostructured thin films for all the samples were used as photoelectrode in the Photoelectrochemical cell for water splitting reaction.Â Maximum photocurrent density of 2.22 mA/cm^2 at 0.95 V/SCE was exhibited by 454 nm thick Ti-Fe_2O_3/Cu_2O bilayered photoelectrode (Fig.1). Increased photocurrent density and enhanced photoconversion efficiency, exhibited by the bilayered thin films may be attributed to improved conductivity and efficient separation of the photogenerated charge carriers at the Ti-Fe_2O_3/Cu_2O interface.Â

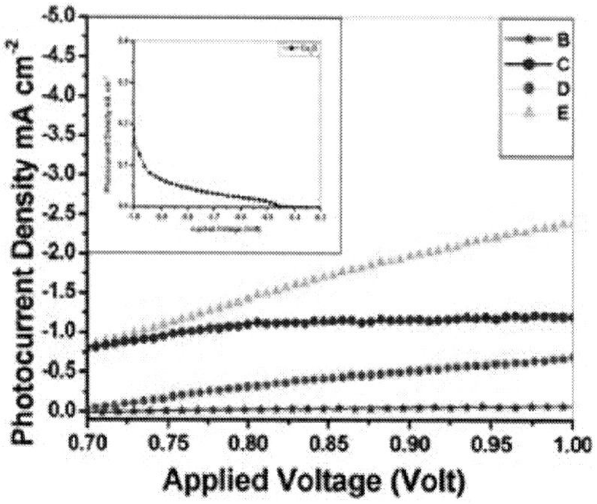

Fig. 1 Photocurrent density vs. Applied potential curve for pure Cu_2O(A), Fe_2O_3(B), Ti ?Fe_2O_3(C), and Fe_2O_3/Cu_2O(D), Ti-Fe_2O_3/Cu_2O(E) bilayered thin films respectively ,under visible light illumination in 0.1M NaOH electrolytic solution using 150W visible light source of irradiation 150 mW cm^{-2} at the position of sample.

Keywords: *Photoelectrochemical, Water splitting, Bilayered,* Cu_2O, Fe_2O_3, *Ti-* Fe_2O_3, *Spray-Pyrolysis*

(499d) Synthesis and Characterization of Mn2O3-ZnO Nanocomposites for Hydrogen Generation Via Photoelectrochemical and Photocatalytic Splitting of Water

Singh, N., Dayalbagh Educational Institute
Choudhary, S., Dayalbagh Educational Institute
Satsangi, V. R., Dayalbagh Educational Institute
Dass, S., Dayalbagh Educational Institute
Shrivastav, R., Dayalbagh Educational Institute
Originally presented on: 11/6/2013 13:27:00 - 13:46:00

Synthesis and Characterization of Mn_2O_3-ZnO Nanocomposites for Hydrogen Generation *via* Photoelectrochemical and Photocatalytic Splitting of Water

Nirupama Singh, Surbhi Choudhary, Vibha R. Satsangi[#], Sahab Dass and Rohit Shrivastav*

Department of Chemistry, [#]Department of Physics & Computer Science,

Dayalbagh Educational Institute, Dayalbagh, Agra-282 005, India

*E-mail: rohitshrivastav_dei@yahoo.co.in

The growing consumption of energy, which is now mainly produced from fossil fuels leads to a further depletion of the limited fossil reserves, increased dependency on fossil fuel imports from often politically unstable counties and furthermore gives rise to an increased concentration of green house gases mainly CO_2 in the atmosphere. The report deals with the synthesis of Mn_2O_3-ZnO nanocomposites, their characterization, and possible use in photoelectrochemical and photocatalytic splitting of water for hydrogen generation. Powdered samples of nanocomposites were prepared by co-precipitation at varying Mn/Zn molar ratios. XRD analysis revealed the evolution of wurtzite ZnO and Mn_2O_3 phases in samples, while optical absorption threshold was found shifted towards higher wavelength with increasing Mn amount. The Zn/Fe molar ratio in nanocomposites was optimized, yielding significantly enhanced photoelectrochemical current and photocatalytic hydrogen yield. Plausible explanations for observed change in photoresponse of samples have been offered.

(499e) High Performance Fluorine Doped Oxygen Evolution Reaction (OER) Electro-Catalysts for PEM Based Water Electrolysis

Kadakia, K., University of Pittsburgh
Datta, M. K., University of Pittsburgh
Velikokhatnyi, O.,
Kumta, P. N., University of Pittsburgh
Originally presented on: 11/6/2013 13:46:00 - 14:05:00

Identification of electro-catalysts containing non noble metal or significantly reduced amounts of expensive noble metals (e. g. RuO_2) is highly desirable. Development of such a catalyst with comparable electrochemical performance to the standard noble metal oxide counterpart for the proton exchange membrane (PEM) based water electrolysis would constitute a major advancement in generation of hydrogen by water electrolysis. We have accordingly developed a two-pronged theoretical first principles and experimental approach to generate a nano-structured solid solution of SnO_2:10wt.% F containing only 20at.% RuO_2 [*e.g.* $(Sn_{0.80}Ru_{0.20})O_2$:10F] that displays a remarkably similar electrochemical activity and moreover, comparable or even much improved electrochemical stability and durability compared to pure the noble metal counterpart, RuO_2. Results of these studies will be presented and discussed.

(499f) Multiwalled Carbon Nanotubes Drive the Activity of Metal@Oxide Core-Shell Catalysts in Modular Nanocomposites

Cargnello, M., University of Pennsylvania

Originally presented on: 11/6/2013 14:05:00 - 14:24:00

Multiwalled Carbon Nanotubes Drive the Activity of Metal@oxide Core-Shell Catalysts in Modular Nanocomposites

M. Cargnello,[1,2] M. Grzelczak,[1] B. Rodriguez-Gonzalez,[3] Z. Syrgiannis,[1] K. Bakhmutsky,[4]

V. La Parola,[5] L. M. Liz-Marzán,[3] R. J. Gorte,[4] M. Prato,[1] and P. Fornasiero[1]

[1] *Department of Chemical and Pharmaceutical Sciences, ICCOM-CNR, Consortium INSTM, University of Trieste, via L. Giorgieri 1, 34127 Trieste, Italy*

[2] *Department of Chemistry, University of Pennsylvania, 231 S. 34th Street, Philadelphia, PA 19104, USA*

[3] *Departamento de Química Física, Universidade de Vigo, 36310 Vigo, Spain*

[4] *Department of Chemical and Biomolecular Engineering, University of Pennsylvania, 311A Towne Building, 220 S. 33rd Street, Philadelphia, PA 19104, USA*

[5] *Istituto per lo Studio dei Materiali Nanostrutturati (ISMN-CNR), Via Ugo La Malfa 153, Palermo I-90146, Italy*

The ability to build hierarchical structures by arranging different building blocks with nanometer-scale precision is one of the most useful aspects of nanotechnology. This concept can produce novel materials with properties that are different from those expected from the simple sum of the individual blocks. Indeed, the interactions between the constituent parts in nanometer-scale ensembles can lead to novel electronic, optical, or catalytic properties not available in the initial building blocks. In the field of heterogeneous catalysis in particular, the interactions between the active phase, supports, and promoters is critical for obtaining high performance materials. These interactions can be both electronic and geometric. Furthermore, the demands in terms of geometry and binding energy of the active sites for different catalytic reactions can be very different.

In this contribution, rational nanostructure manipulation has been used to prepare nanocomposites in which multiwalled carbon nanotubes (MWCNTs) were embedded inside mesoporous layers of oxides (TiO2, ZrO2, or CeO2), which in turn contained dispersed metal nanoparticles (Pd or Pt). We show that the MWCNTs induce the crystallization of the oxide layer at room temperature and that the mesoporous oxide shell allows the particles to be accessible for catalytic reactions. In contrast to samples prepared in the absence of MWCNTs, both the activity and the stability of core-shell catalysts is largely enhanced, resulting in nanocomposites with remarkable performance for the water-gas-shift reaction, photocatalytic reforming of methanol, and Suzuki coupling. The modular approach shown here demonstrates that high-performance catalytic materials can be obtained through the precise organization of nanoscale building blocks.

(499g) Separation of Sulphur Dioxide and Oxygen in Thermochemical Hydrogen Production

Elder, R. H., University of Sheffield
He, G., University of Sheffield
Cumming, D. J.,
Allen, R. W. K., University of Sheffield

Originally presented on: 11/6/2013 14:24:00 - 14:43:00

Dense Yttria-stabilised-zirconia (YSZ) membranes have been shown to have great potential for separating SO_2 and O_2 in the sulphur family of thermochemical cycles for hydrogen production. Use of a catalytic YSZ membrane reactor in the sulphuric acid decomposition stage of these cycles could drive the reaction forward and significantly increase the yield of SO_2 and O_2. Whilst SO_2/O_2 separation has been shown to be feasible at temperatures between 750°C and 950°C, significant performance degradation is seen over time. This performance degradation is shown to be due to sulphur poisoning (from both sulphur adsorption and sulphide formation) of the triple phase boundaries. A similar effect is seen in solid oxide fuel cells, where ppm concentrations of sulphur severely diminish cell performance.

Here we investigate the effect of voltage pulsing on membrane performance recovery. Significant performance recovery is seen after the application of a short voltage pulse. The optimum pulse duration and magnitude were found to be 0.08 s and 10 V respectively. The lower the magnitude of voltage pulse, the longer the duration required to achieve maximum recovery. Impedance spectroscopy and SEM results are presented, showing the contribution of different resistances and the effect of the voltage pulsing on the membrane cell.

Consideration of oxygen reduction mechanisms at the triple phase boundary shows that as voltage across the membrane increases, the rate controlling step changes from oxygen adsorption and surface diffusion, to oxygen availability in the feed. The higher voltage pulses are beneficial to performance recovery as polarisation resistance and charge transfer resistance decrease due to the removal of sulphur from the surface. Pulses that have too great a magnitude, or are too long, dramatically promote oxygen reduction and lead to concentration polarisation due to lack of oxygen in the feed.

Longer term testing (20 hours) showed that repeated, regular voltage pulsing can significantly improve membrane performance compared to those that receive no voltage pulsing.

(499h) Chemisorption, Physisorption, and Hysteresis of Hydrogen On Carbon Nanotubes

Barghi, S., University of Southern California

Tsotsis, T. T., University of Southern California

Sahimi, M., University of Southern California

Originally presented on: 11/6/2013 14:43:00 - 15:02:00

We used a magnetic suspension balance, coupled with a residual gas analyzer, to investigate hydrogen adsorption/desorption on carbon nanotubes (CNTs) for pressures up to 100 bars at 25 °C. Such a combination of gravimetric and mass analysis systems, used for the first time, is capable of overcoming the shortcomings of the common hydrogen adsorption measurement techniques. The most crucial drawbacks of the conventional measurement methods are as follows. (i) The inability to distinguish between hydrogen adsorption and chemisorptions; (ii) the relatively large errors caused by the leakage of hydrogen out of the system, and (iii) the difficulty of measuring hydrogen uptake capacity under operating conditions that differ from those of hydrogen storage systems used in practice.

In the experiments multi-walled carbon nanotubes (MWCNTs) were exposed to hydrogen, and were subsequently kept under vacuum for 2 consecutive cycles, in order to study hydrogen adsorption/desorption in the nanostructured materials. The difference between the first and the second hydrogenation cycles was in the initial preparation of the MWCNTs. For the first cycle, the MWCNTs were first degasified at an elevated temperatures (120 °C). For the second cycle the MWCNTs were not degasified at the elevated temperatures and, therefore, hydrogen residue from the first adsorption/desorption cycle was still adsorbed on the MWCNTs at the beginning of the second cycle.

The first hydrogenation cycle was started after pretreatment of the CNT at the elevated temperature. Hydrogen adsorption kinetics for the cycle indicates that the adsorption equilibrium occurs within 4 - 5 hours. Careful examination of the recorded adsorption kinetics revealed the fascinating fact that a considerable part of hydrogen adsorption in the MWCNTs takes place *instantaneously*. The measured equilibrium data revealed some hysteresis for hydrogen desorption in this cycle, which does not vanish, even when the system's pressure approaches zero. This indicates that hydrogen adsorption in the first hydrogenation cycle is a combination of physisorption and chemisorption.

After the first cycle was finished, the CNTs were exposed to hydrogen for the second time without any pretreatments at the elevated temperature. Measurements indicated no hysteresis for during the second cycle. Therefore, only physisorption took place during the second cycle. Comparison between the first and the second hydrogen isotherms, together with the measured hydrogen adsorption kinetics, made it possible to quantitatively distinguish between the physisorption and chemisorption portions of hydrogen adsorption in the MWCNTs. According to our data, for pressures less than 20 bars physisorption and chemisorption are both important, with the former becoming the dominant mechanism as the pressure increases. To the best of our knowledge, this is the first time that hydrogen physisorption and chemisorption on carbon nanotubes have been measured separately.

(558a) Thermoelectric Properties of Ultra-Long PbSe Hollow Nanofibers

Zhang, M., University of California, Riverside
Park, H., University of California-Riverside
Jung, H., University of California, Riverside
Kim, J.,
Kim, S.,
Lim, J. H., Korea Institute of Materials Science
Choa, Y. H.,
Myung, N., University of California, Riverside
Originally presented on: 11/6/2013 15:15:00 - 15:45:00

Lead selenide (PbSe), a sister material of lead telluride (PbTe), has been less focused on for thermoelectric (TE) applications due to its smaller band gap energy and an expected higher thermal conductivity. Recently, however, an additional notion has been placed to the material, since a TE figure of merit (ZT) up to 1.2 at 850 K was experimentally achieved, approaching the theoretical value of 1.6. In addition, significant progress has been made to enhance the Seebeck coefficient while diminishing the thermal conductivity by using low dimensional PbSe. However, even though a thermal power 3 times greater than that of bulk in PbSe quantum dot superlattices has been reported, only a low power factor was extracted attributing to its high electrical resistivity. One-dimensional (1-D) PbSe, on the other hand, shows a better electrical conductivity with a significant decreased thermal conductivity, but the prevailing difficulty and high cost in a device fabrication hinders its application in a real TE apparatus. Therefore, low dimensional PbSe with superior TE performance is highly demanded, through a cost-effective fabrication process.

While several techniques exist for creating 1-D PbSe nanostructures, electrospinning (ES) has emerged as the most versatile, scalable and cost-effective method to synthesize ultra-long nanofibers (NFs) with controlled diameter and composition. Although various NFs have been synthesized using this technique, limited work has been reported on chalcogenide NFs synthesis due to the incompatibility of the precursors. Therefore, synthesis of PbSe was conducted through ES with the addition of a galvanic displacement reaction (GDR); this is a spontaneous electrochemical process driven by the difference in materials' redox potentials. In this approach, ES is exploited to fabricate ultra-long sacrificial NFs with controlled dimensions, morphology and crystal structures, providing a large materials' data base to tune electrode potentials thereby imparting control over the composition and shape of the nanostructures evolved during GDR. Synthesized samples are designed to be in a bulk scale for easy device fabrication, but reserve the superior properties form nanoscale materials.

In this work, we demonstrated high scalability and cost-effective nanofabrication to synthesize ultra-long hollow PbSe nanofibers by combining ES and a GDR. Control over the diameter, wall-thickness, morphology, composition and crystallinity of the NFs was achieved by tuning the shape and dimension of the sacrificial materials as well as GDR conditions. Temperature dependent electrical resistivity and Seebeck coefficient measurement were performed for the TE properties characterization. Preliminary result showed an excellent Seebeck coefficient of 525 μV/K and a power factor of 0.43 $\mu W/K^2m$ at 26 °C in the PbSe hollow NF mat.

(558b) Thermal Transport in Nanocrystal Arrays and Self Assembled Monolayers

Ong, W. L., Carnegie Mellon University
Rupich, S., University of Chicago
Majumdar, S., Carnegie Mellon University
Talapin, D. V., University of Chicago
McGaughey, A. J. H., Carnegie Mellon University
Malen, J. A., Carnegie Mellon University

Originally presented on: 11/6/2013 15:45:00 - 16:15:00

The thermal conductivity of nanocrystal arrays (NCAs) is found to be tunable with chemistry and the nanocrystal diameter. Nanocrystals (inorganic crystalline cores 2-10 nm in size encapsulated in organic monolayers) self-assemble from colloidal suspensions into close-packed 3D NCA films. These films exhibit tunable electronic and optical properties that can meet various demands in energy conversion and optoelectronic applications. When coupled with scalable solution-based fabrication processes, these properties make NCAs a cheap and versatile replacement for conventional single crystal semi-conductors and nanostructured materials.

Though electronic transport in NCAs has been studied extensively, little is known, about the nature of thermal transport in these complicated materials. We show that the thermal conductivity of NCAs shows independence from the bulk core thermal conductivity, yet is tunable from 0.1-0.3 W/m-K by increasing the nanocrystal diameter. Effective medium approximations (EMA) modified by incorporating a fintite thermal conductance at the core ligand interface correctly captures the measured diameter-dependent magnitudes and trends. Changing the chemistry of the nanocrystal (i.e., different core and ligand species) enables further control for manipulating the NCA thermal conductivity. To learn more about thermal transport at this interface we have directly studied the thermal conductance of a series of self assembled monolayer (SAM) junctions. By considering various metal-SAM-metal junctions we find that the interface conductance is influenced by both the SAM-metal bond and the overlap of their vibrational states. Molecular dynamics simulations of NCAs and SAMs help us to sort out these effects and how they more generally apply at the organic-inorganic interface.

(558c) Thermoelectric Nanocomposites of Layered Chalogenide Bi2(Te/Se)3 Nanoplateletes and Their Interfacial Effects

Soni, A.,
Xiong, Q.,

Originally presented on: 11/6/2013 16:15:00 - 16:45:00

Thermoelectric nanocomposites of layered chalogenide $Bi_2(Te/Se)_3$ nanoplateletes and their interfacial effects

Ajay Soni[1,2] and Qihua Xiong [*,2,3]

[1] School of Basic Sciences, Indian Institute of Technology Mandi, Mandi, Himachal Pradesh 175001, India

[2] Division of Physics and Applied Physics, School of Physical and Mathematical Sciences, Nanyang Technological University, Singapore 637371, Singapore

[3]Division of Microelectronics, School of Electrical and Electronic Engineering, Nanyang Technological University, Singapore 639798, Singapore

* To whom the correspondence should be addressed. Email address: Qihua@ntu.edu.sg,

Recently the interests in layered chalcogenide materials have grown very rapidly. The compounds of chalcogenides (S, Te and Se) with Bi and Sb are of interests for topological insulators, thermoelectric (TE) applications, while the compounds with transition metals are of interests for optoelectronic applications. The observed fascinating physical properties are arising mainly due to the anisotropic crystal structures, weak van der Waals stacking of layer and quantum confinement effects in atomic thin layers. Historically, chalcogenide materials have been known as good TE materials, especially the family of $(Bi/Sb)_2(Te/Se)_3$ alloys which showed a competitive performances for room temperature TE applications.[1] The efficiency of the TE material is quantified by a dimensionless TE figure of merit, $ZT = S^2s/k$, where S is the Seebeck coefficient, s is the electrical conductivity and k is the thermal conductivity. All three of the physical parameters are strongly dependent on the number of charge carriers (doping) and the arrangements of interfaces/grain boundaries. Thus the interplay of the three physical parameters leads to a very complex situation for engineering of charge and heat carriers in TE materials. For the new class of thermoelectric (TE) nanocomposites,[1-3] while the energy filtering by interfaces and grain boundaries is playing a vital role for enhancing the performances of TE materials, the thermodynamical instability of grain boundaries limits the application range and performance reproducibility.[1] To highlight this theme, we will present our work on the enhanced thermoelectric properties of chemically synthesized 2D n-type $Bi_2(Te/Se)_3$ nanoplatelet composites[2] and the effects of the thermodynamic environment during spark plasma sintering on the TE performance of $Bi_2Te_{2.7}Se_{0.3}$ nanoplatelet composites.[3] Our studies emphasize on a unique coupling charge carriers and phonons and a subtle understanding of the interface crystallization, grain growth and energy filtering from the thermoelectric interfaces.

1. Dresselhaus, M. S., *Adv. Mater.* **2007,** *19*, 1043-1053.

2. Soni, A.; Yanyuan, Z.; Ligen, Y.; Aik, M. K. K.; Dresselhaus, M. S.; Xiong, Q., *Nano Letters* **2012,** *12* (3), 1203-1209.

3. Soni, A.; Yiqiang, S.; Ming, Y.; Yanyuan, Z.; Ligen, Y.; Xiao, H.; Zhili, D.; Aik, M. K. K.; Dresselhaus, M. S.; Xiong, Q., **2012,** 12 (8) 4305.

(558d) Advanced Nanocomposite Fibers for Thermoelectric Energy Harvesting and Motion Sensing

Wu, Y., Purdue University
Finefrock, S., Purdue University
Originally presented on: 11/6/2013 16:45:00 - 17:15:00

In this presentation, we will discuss nanocrystal coated fiber based devices for thermoelectric energy harvesting from waste heat and the conversion of air movement (including wind as well as air flow created by motion). In addition, we will show that these devices are capable to detect an extremely small temeprature fluctuation created by motion, which could find many defense and civilian applications. We will also describe our investigations on the thermal conductivity in the axial direction of a single glass fiber coated with lead telluride (PbTe) nanocrystals using the self-heated 3ω method particularly at low frequency. While prior 3ω measurements have only been demonstrated for high thermal conductivity materials, the present work demonstrates the suitability of the 3ω method for PbTe nanocrystal coated glass fibers where the low thermal conductivity and high aspect ratio result in a significant thermal radiation effect. We simulated the experiment using a finite-difference method that corrects the thermal radiation effect and extract the thermal conductivity of a single glass fiber coated by PbTe nanocrystals to be 0.32-0.46 W/mK in the temperature range 260 K-300 K. This process of including both an experimental approach and numerical modeling could be used to study the intrinsic thermal conductivity of other one-dimensional nanowire or nanotube samples with low thermal conductivity at the single wire/tube level, thus significantly broadening the application of the 3ω measurement technique.

(558e) Development of a Nanocatalytic Microcombustor Power Device

McNally, D.,
Bakrania, S.,
Originally presented on: 11/6/2013 17:15:00 - 17:33:00

With the rising demand for portable power devices, the production of energy through micro-scale combustion of hydrocarbon fuels is becoming increasingly attractive. Microcombustors coupled with thermoelectric devices have the potential to provide a prolonged power source for portable devices as opposed to conventional batteries due to significantly higher energy densities. Recharging then can be easily achieved by simply adding more fuel for such integrated devices, making them desirable for remote operations. Current research is focused on the design of microreactors to sustain combustion and manage temperature gradients within. Previously, our work has successfully demonstrated room temperature ignition and control over catalytic activity of methanol-air mixtures with platinum nanoparticles. For this work, we have developed a microcombustor power generation device to explore the potential application of nanocatalytic combustion. Commercial thermoelectric generators were used to convert the temperature gradient created by combustion directly to electrical power. The overall performance of the device, together with the temperature characterization of the device, will be reported. Key parameters that influence performance and contribute to the modeling efforts were also identified. This work indicates a potential for a highly miniaturized design for such devices. At the same time, our own catalytic combustion research continues to drive the future design of an alternative power source to conventional battery technology.

(558f) Templated Fabrication and Characterization of Thermoelectric Nanowire Arrays - Toward Power-Dense and Efficient Devices

Crosser, L. A., Mainstream Engineering Corporation
Chester, G. E., Mainstream Engineering Corporation
Yelvington, P. E., Mainstream Engineering Corporation
Hill, J. J., Mainstream Engineering Corporation
Originally presented on: 11/6/2013 17:33:00 - 17:51:00

Efficient and power-dense thermoelectric devices would offer significant advantages over other forms of thermal energy conversion and control technologies. A promising route toward enhanced efficiency is through nanostructuring thermoelectric materials, the benefits of which have been predicted and demonstrated by many. However, cost effective assembly that preserves the performance enhancement of the individual nanostructures remains the dominant engineering challenge. Specifically, much attention has been focused on high aspect ratio structures such as nanowires, as they can directionally transport charge carriers and disrupt heat transport without the presence of grain boundaries or excessive interfacial sites. We are developing fabrication and characterization methods for the templated growth and assembly of thermoelectric nanostructures, including advanced low-energy nanofabrication techniques designed to both increase nanostructure anisotropy and enhance aerial power density compared to traditional templated approaches. Empirical and theoretical aspects of the fabrication methods will be highlighted with a focus on maintaining high power density (assembly density) and high efficiency (structure properties & confinement). We will discuss a self-assembly coating procedure which increases radial confinement in nanoporous oxide templates to produce densely arrayed and ultra-low diameter thermoelectric nanowires. Furthermore, we will discuss a method to supplant the loss in aerial power density introduced by the oxide template and its enhancement with respect to overall performance. Thermoelectric property measurement of thin films and anisotropic structures is non-trivial as thermal and electrical contact resistances significantly contribute to the measurement. We will briefly introduce an apparatus that measures thermopower as well as thermal and electrical conductivity designed specifically for thin film or anisotropic structures.

(558g) Discovering Materials With Ultra-Low Work Functions for Thermionics Energy Conversion

Chou, S. H., Stanford University
Voss, J., Stanford University
Vojvodic, A., SLAC National Accelerator Laboratory
Howe, R. T., Stanford University
Abild-Pedersen, F., Technical University of Denmark
Originally presented on: 11/6/2013 17:51:00 - 18:09:00

The work function is one of the most fundamental properties of a metallic surface, important in determining the material's applicability as a contact electrode [1, 2] or electron emitter [3]. Surfaces with ultra-low work functions could therefore improve technologies requiring precise control of contact barriers such as organic and printed electronics [2] or a variety of devices based on electron emission, ranging from fluorescent light bulbs [4, 5] to THz sources [6], thermionic energy converters (TECs) [7], and the photon-enhanced thermionic emission converters (PETEC) [8]. For TEC and PETEC, in particular, discovery of thermally stable materials with work functions of less than 1 eV would allow thermionic conversion of high-temperature (> 500 °C) heat or solar radiation directly to electricity with efficiencies exceeding 50%.

We present a model based on adsorbate atomic orbital interactions which offers new insights into the interfacial phenomena that control the work function [9]. In addition to the model, we will discuss *ab-initio* studies of surfaces with alloyed alkali metal and alkali-earth metal oxides. These adsorbed compound structures can enhance the emissivity of refractory metal substrates commonly used in current electron-emissive devices. By visualizing the electrostatic charge distributions, surface dipole formations, and the corresponding changes in the work function, we can devise design rules for low-work function materials to be incorporated into high-efficiency thermionic energy conversion devices.

[1] Lindell L, Burquel A, Jakobsson F L E, Lemaur V, Berggren M, Lazzaroni R, Cornil J, Salaneck W R and Crispin X 2006 Chem. Mater. **18**, 4246

[2] Zhou Y *et al* 2012 Science **336**, 327

[3] Yamamoto S 2006 Rep. Prog. Phys. **69**, 181–232

[4] Wada M, Gotoh S and Kurumada S 2009 J. Plasma Fusion Res. Series **8**, 1366

[5] Watanabe S, Watanabe T, Ito K, Miyakawa N, Ito S, Hosono H and Mikoshiba S 2011 Sci. Technol. Adv. Mater. **12**, 034410

[6] Snapp J P, Lee J-H, Provine J, Bargatin I, Maboudian R, Lee T H and Howe R T 2012 Sidewall silicon carbide emitters for terahertz vacuum electronics Solid-State Sensor and Actuator Workshop (San Diego, CA, June)

[7] Lee J-H, Bargatin I, Provine J, Clay W, Schwede J, Liu F, Maboudian R, Melosh N, Shen Z-X and Howe R T 2009 Tech. Dig. Power MEMS

[8] Schwede J W *et al* 2010 Nature Mater. **9**, 762

[9] Chou S H, Voss J, Bargatin I, Vojvodic A, Howe R T and Abild-Pedersen F 2012 J. Phys.: Condens. Matter **24**, 445007

(630a) Silicon Nanostructures As Efficient Thermoelectric Materials

Yang, P., Unversity of California, Berkeley
Lim, J.,
Originally presented on: 11/7/2013 8:30:00 - 9:00:00

It has been recently reported that thermal conductivity greatly decreases in single crystalline silicon nanostructures without much affecting electrical properties. Although it is qualitatively demonstrated that surface roughness in nanowires and nanohole arrays in silicon membranes (holey silicon) play a significant role in reducing thermal conductivity, the underlying reasons remain unknown and warrant quantitative studies and analysis. In this report, silicon nanowires were controllably roughened and the surface roughness near the length scale of thermal phonon wavelength is shown to interfere with phonon transport below the Casimir limit. Furthermore, a platform to simultaneously measure thermal conductivity, electrical conductivity, and the Seebeck coefficient for holey silicon has been developed to extract ZT directly from a single device. Decoupling of electron and phonon transport associated with their mean free path has been studied at phonon glass electron crystal regime. Lastly, cross-plane thermal conductivity measurement of holey silicon unveiled the multi directional phonon-glass property of holey silicon.

(630b) Toward Physical Models of Thermoelectric Transport Rules At the Organic-Inorganic Interface

Urban, J., Lawrence Berkeley National Laboratories
Originally presented on: 11/7/2013 9:00:00 - 9:30:00

The thermoelectric properties of solution-processed conducting polymer/nanoparticle composite films have been a focus of interest for many groups due to their unique transport mechanisms, enabling new modalities of transport which can surpass common optimization dilemmas in inorganic thermoelectrics. However, many interesting reports have lacked a theoretical framework to guide materials design. Here, I present a few organic/inorganic thermoelectric model systems studied as a function of nanoparticle loading and chemical doping. We use these to develop phenomenological frameworks to guide engineering of this novel material class.

Here I will discuss two different hybrid systems, and the tunability and optimization of their thermoelectric performance: 1.) Poly(3,4 ethylenedioxythiophene):poly(styrenesulfonate) (PEDOT:PSS) and Tellurium nanowire devices, and Polyaniline and nanoparticle-Titania devices. For each of these systems, interesting non-monotonic electrical conductivity is observed at intermediate loadings, suggesting non-effective medium behavior, in contrast to the values of the Seebeck coefficient increase linearly with mass fraction of inorganic. These trends in Seebeck coefficient and electrical conductivity lead to an optimized power factor in the composites that exceed that of the individual components. The surprising electrical conductivity behavior can be explained with a mixing model where carrier transport is primarily through a highly conductive volume of polymer that exists at the nanoparticle-polymer interface. We have developed a general phenomenological model for these systems suggesting that this increase in conductivity results from a structural modification of the polymer, and elucidates a novel route for optimizing transport in nanoscale organic/inorganic composites.

(630c) Thermal and Thermoelectric Transport in Semiconductor Nanowires

Chen, R., University of California, Berkeley

Originally presented on: 11/7/2013 9:30:00 - 10:00:00

Semiconductor nanostructures are ubiquitous in electronic devices. Thermal transport plays an important and sometimes pivotal role in the performance of these devices. For instance, thermal coupling between nano-scale transistors and the surrounding materials would dictate the heat dissipation of the transistors. Thermal conductivity is an important parameter to engineer to enhance energy conversion efficiency of thermoelectric devices. During the past decade, we have witnessed tremendous efforts devoted to the understanding and manipulation of thermal transport properties in semiconductor nanostructures, including thin films, superlattices, nanowires and nano-composites. However, most of these studies have been focused on the classical size regime where the characteristic sizes of nanostructures are greater than dominant phonon wavelength. As the size of nanostructures keeps shrinking and becomes comparable to phonon wavelength, the quantum confinement effect would emerge. Exact picture of phonon transport in this new regime, however, is not fully explored. In this presentation, we use semiconductor nanowires with ultra-small diameters as a model system to explore the quantum confinement effect on thermal and thermoelectric transport. We will present experimental techniques we have developed recently to probe thermal and thermoelectric transport properties in thin nanowires and the new understanding obtained from these experimental studies.

(630d) Nanowire Heterostructure-Based Thermoelectric

Wu, Y., Purdue University
Yang, H., Purdue University
Fang, H.,
Originally presented on: 11/7/2013 10:00:00 - 10:20:00

In this presentation, we will describe our latest progress on the general, reproducible, and scalable synthesis of nanowire heterostructures for potential thermoelectric energy harvesting for low grade heat recovery. Particularly, through a multiple step solution phase process, we can mass produce telluride based thermoelectric nanowire heterostructures with embedded nanoparticles or nanoplates at two ends. In certain cases, significantly improved performance compared to bulk materials has been observed due to the contribution of reduced thermal conductivity through interface phonon scattering and enhanced Seebeck coefficient through energy filtering. The synthesis method could be used to produce nanowires up to kilogram level using our large batch reactor while still maintaining a narrow diameter/length/performance distribution, which could open the door for the real applications of nanostructured thermoelectric materials. These nanowire heterostructures also provide novel platforms in a reproducible way to study new theories in electrical and thermal transport at nanoscale interface inculiding quantum confinement, energy filtering, and phonon scattering.

(630e) Thermoelectric Properties of Bulk Pellets of Unfunctionalized and Functionalized Zn3P2 Nanowires

Brockway, L., Texas A&M University
Vasiraju, V.,
Vaddiraju, S., Texas A&M University
Originally presented on: 11/7/2013 10:20:00 - 10:40:00

Zinc phosphide (Zn_3P_2) nanowire powders, both unfunctionalized and surface-functionalized with organic molecules to passivate the surfaces, were mass produced using reactive vapor transport of phosphorus onto zinc foils in a hot-walled chemical vapor deposition chamber, and hot-pressed into dense pellets for the characterization of their thermoelectric performance. The intent here is to explore the possibility of fabricating thermoelectric modules using inexpensive compound semiconductors. Contrary to expectations, the hot-pressing did not alter the morphology or the dimensions of the nanowires in the pellets, despite the fact that high packing densities (>98%) were achieved in the pellets. It was observed that enhanced flexibility observed in thin nanowires allows for their assembly into highly dense pellets, without any morphological alterations. The elastic nature of the Zn_3P_2 nanowires with diameters less than 50 nm was confirmed using *in-situ* transmission electron microscopy (TEM) studies.

Overall, this experimentation indicated that a 30% reduction in thermal conductivity relative to the bulk is feasible by synthesizing Zn_3P_2 in nanowire format. At the same time, this procedure did not adversely affect the electrical conductivity. In fact, prevention of surface oxidation by functionalizing the nanowires with 1,4-benzenedithiol before their assembly into pellets led to an overall increase in the electrical conductivity of the nanowire pellets. These measurements indicate that it is possible to enhance the thermoelectric performance of materials by synthesizing them in nanowire format. The thermoelectric performance of both unfunctionalized and functionalized Zn_3P_2 nanowire pellets, along with those obtained using by doping them with copper, will be discussed in detail in this presentation.

(630f) Combustion Synthesis of Thermoelectric Oxide Powders

Lin, S., Lamar University
Duruewuru, P., Lamar University
Bonura, J., Lamar University
Originally presented on: 11/7/2013 10:40:00 - 11:00:00

Doped calcium cobaltate powders were prepared by Self-propagating High-temperature Synthesis (SHS). Reactant powders are consolidated to a pellet and ignited by an external heat source. The heat released from the reaction is sufficient to sustain the movement of the reaction front to convert the reactants into the product. TG-DSC analysis was performed to establish the reaction mechanism. Thermocouples and an IR camera were used to measure the temperature history at the pellet centerline and pellet surface during the combustion to determine the velocity of combustion front movement. Products were characterized by XRD for phase purity. Synthesized powders were hot-pressed to form dense pellets and characterized for their thermoelectric properties. Our results show that SHS can economically produce thermoelectric oxides with high ZT values and the thermal conductivity of calcium cobaltate decreases when silver is doped to replace calcium.

(684a) Time Resolved Ion Dynamics in Carbon Supercapacitor Electrodes Using In Situ Infrared Spectroscopy

Richey, F.,
Dyatkin, B.,
Gogotsi, Y.,
Elabd, Y. A., Drexel University
Originally presented on: 11/7/2013 12:30:00 - 12:46:00

Ion transport in energy storage devices (e.g., batteries, supercapacitors) is the fundamental process that dictates charge/discharge rates. However, there are few experimental techniques that can measure molecular level ion transport in response to an input voltage. In this study, we apply a novel *in situ* infrared spectroscopy measurement technique to investigate ion transport in carbon electrodes of electrochemical double layer supercapacitors. Ion transport properties of porous nano-sized carbide derived carbons (CDC) were compared to non-porous carbon onions for both pure ionic liquid electrolytes and mixtures of propylene carbonate and ionic liquid.

(684b) Tailoring the Properties of Polymer- and Silica-Supported Ionogel Electrolytes for Energy Storage Applications

Panzer, M. J., Tufts University

Originally presented on: 11/7/2013 12:46:00 - 13:02:00

A burgeoning demand for portable electronic devices, as well as an expanding global portfolio of intermittent electricity generation sources both highlight the critical need to develop reliable, safe, and efficient energy storage technologies. Recently, our group has been advancing two parallel approaches to form solid electrolyte films based on room temperature ionic liquids (ionogels) for supercapacitor applications.[1,2] The first method utilizes free radical-initiated polymerization inside an ionic liquid to produce a composite gel, while the second approach employs an acid-catalyzed sol gel reaction network to create an inorganic oxide support for the ionic liquid *in situ*. Ionogels are inherently safer than widely used organic solvent-based electrolytes due to their nonvolatility and no-leak nature. Optimization of ionic conductivity for fast charging and discharging in supercapacitor structures within the constraint of maintaining gel mechanical integrity will be discussed. In the best-case scenario, both polymer- and silica-supported ionogels retain nearly identical electrical properties as compared to their constituent neat ionic liquids, while also providing sufficient flexibility and robustness for intended future applications. Our recent work exploring the recovery and recyclability of ionic liquids from silica-supported ionogels using a green solvent-based process with an extremely low energy requirement will also be presented. [1] A.F. Visentin, M.J. Panzer, ACS Appl. Mater. Interfaces 2012, 4, 2836. [2] A.I. Horowitz, M.J. Panzer, J. Mater. Chem. 2012, 22, 16534.

(684c) Pseudocapacitance Dominant Aqueous Asymmetric Supercapacitor Based On Manganese Oxide Nanoflowers

Mao, Y., Washington State University
Li, Q., Institute of Process Engineering, Chinese Academy of Sciences
Originally presented on: 11/7/2013 13:02:00 - 13:18:00

Electrochemical capacitors (ECs) are promising power sources for portable electronics and hybrid electric vehicles. To solve the poor ionic conductivity, intrinsic inflammability and toxicity issues of current ECs incorporating organic electrolytes, aqueous electrolyte-based asymmetric supercapacitors (ASCs) have been attracting intensive attention recently by applying a battery-like Faradic cathode and a capacitive anode. In this presentation, we first report a facile synthesis of highly monodisperse nanostructured MnO_2 flowers with averaged diameter of ~25 nm via redox reaction between $KMnO_4$ and N-Methyl-2-pyrrolidone under mild conditions. Then neutral aqueous ASCs were assembled using these nanostructured MnO_2 as cathode and functional carbon nanotubes (CNTs) as anode. These ASCs exhibits an operational window as high as 2.0 V, a highest energy density of 29.2 Wh kg^{-1} at a power density of 105 W g^{-1} and competitive rate performance. The electrochemical performances, especially the rate capacity, of these ASC devices have been improved by the monodisperse and fine MnO_2 nanoflowers and intertwined CNT networks with complementary potential windows. This novel nanoarchitecture offers more accessible sites for the surface electrochemistry/adsorption and short paths for ionic diffusion/percolation, and high ionic conductivity and safety by the selected neutral aqueous electrolyte. This study demonstrated a robust synthesis route for MnO_2 nanoflowers as the cathode material, but also synergistic effects of ASC components for their practical applications.

(684i) Plating/Stripping Performance of Different Substrates in Magnesium Borohydride Solution
Saraidaridis, J. D., University of Michigan
Vardar, G., University of Michigan
Sleightholme, A. E. S., University of Michigan
Monroe, C. W., University of Michigan
Siegel, D. J., University of Michigan
Originally presented on: 11/7/2013 13:18:00 - 13:34:00

Plating/Stripping Performance of Different Substrates in Magnesium Borohydride Solution

James D. Saraidaridis[1], Gulin Vardar[2], Alice E. S. Sleightholme[1], Donald J. Siegel[3], Charles W. Monroe[1]

[1]*Department of Chemical Engineering, University of Michigan*

[2]*Department of Materials Science and Engineering, University of Michigan*

[3]*Department of Mechanical Engineering, University of Michigan*

2300 Hayward St

Ann Arbor, MI 48109

High theoretical energy densities have attracted many researchers to metal-oxygen batteries in search of post-Li-ion technologies. The $Mg\text{-}O_2$ chemistry offers higher theoretical volumetric energy densities than popular $Li\text{-}O_2$ chemistry [1][2], which is important in constrained-volume applications like electric vehicles. A magnesium system has large potentials associated with oxygen reactions forming MgO (2.95V vs. Mg/Mg^{2+}) and MgO_2 (2.91V vs. Mg/Mg^{2+}). The largest hurdle to demonstrating rechargeable $Mg\text{-}O_2$ batteries remains the development of suitable electrolytes that allow plating/stripping of Mg metal, prevent side reactions at potentials dictated by the plating and oxygen reactions, and are chemically inert to O_2.

Researchers recently demonstrated the use of magnesium borohydride ($Mg(BH_4)_2$) in DME as an electrolyte in a Mg-ion battery [3]. We sought to explore the performance of this electrolyte with various electrode substrates. The use of microelectrodes and a consistent dimensionless scan rate allows a fair comparison between electrodes of different material and radii. While no substrate was found stable across the relevant potentials for $Mg\text{-}O_2$ reaction, there is a clearly demonstrated difference in the substrate activity for the Mg plating/stripping reaction.

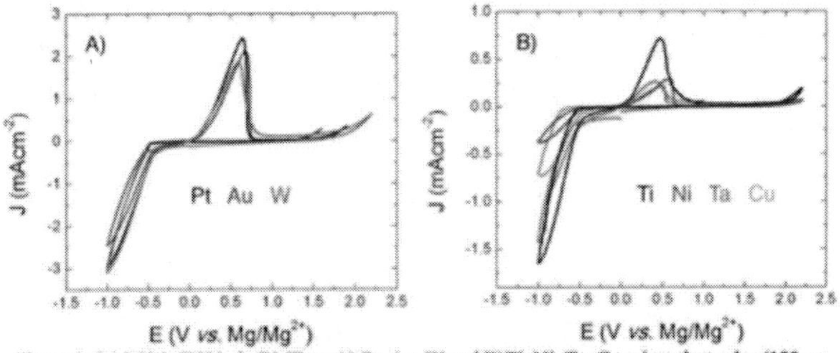

Figure 1: 0.1 M $Mg(BH_4)_2$ in DME on A) Pt, Au, W and B)Ti, Ni, Ta, Cu microelectrodes (100μm and 127 μm), Ar atmosphere <1pp, O_2, H_2O, constant dimensionless scan rate: (A)(v)=0.062mVmm²s⁻¹

[1] C. Zu, H. Li. (2011). *Energy & Environmental Science.* **4:** 2614-2624.

[2] K. Abraham, Z. Jiang. *J. Electrochem. Soc. 143: 1-5.*

[3] Mohtadi, M. Matsui, T. Arthur, S. Hwang. *Angewandte Chemie.* **51**: 9780-9783.

(684e) Vacnt-Based All Solid-State Flexible Micro-Supercapacitors

Hsia, B., University of California, Berkeley
Marschewski, J., University of California, Berkeley
Wang, S., University of California, Berkeley
In, J. B., University of California, Berkeley
Carraro, C., University of California, Berkeley
Poulikakos, D., ETH Zurich
Grigoropoulos, C. P., University of California, Berkeley
Maboudian, R., University of California, Berkeley
Originally presented on: 11/7/2013 13:50:00 - 14:06:00

Flexible microscale electronic devices are garnering increased research and industrial attention due to possible applications in small form factor consumer electronics, bendable displays, and wearable devices. For autonomous, wireless operation, flexible energy storage devices must also be fabricated to power device circuits and communications. Maintaining a planar architecture for this microscale energy storage could simplify integration, reduce cost, and minimize the footprint of such devices. In particular, micro-supercapacitors present an attractive energy storage solution due to their high power density and long cycle lifetime relative to batteries. Supercapacitors, which rely on electrostatic charge storage at the electrode/electrolyte interface, rather than electrochemical energy storage in the electrode bulk as in batteries, would likely require less maintenance than similar battery technologies, particularly for frequent loading applications such as wireless data transmission. One fabrication and integration challenge for microscale flexible supercapacitors and batteries is the encapsulation of liquid electrolytes, which present safety and reliability concerns due to possible leakage. A flexible solid-state solution, therefore, would be extremely attractive if it could alleviate these difficulties without sacrificing mechanical flexibility.

We demonstrate a facile and scalable technique for fabrication of flexible micro-supercapacitors using vertically aligned carbon nanotube (VACNT) electrodes and an ionic-liquid based gel (ionogel) electrolyte. VACNTs present good electronic conductivity as well as fast ionic transport due to their alignment and are hence a prime candidate for supercapacitor electrodes. The VACNTs are easily patterned and transferred onto flexible substrates in a single step, via a laser assisted process. In conjunction with a flexible, ion conducting ionogel, planar micro-supercapacitor devices are constructed in an interdigitated electrode architecture. Fabricated devices are tested via a variety of electrochemical characterization techniques including cyclic voltammetry, galvanostatic discharge, and AC impedance spectroscopy. Mechanical fatigue tests demonstrate the good flexibility of the device. This novel combination of electrode and electrolyte materials, fabricated via a facile, scalable process, demonstrates an attractive avenue for integrated all solid-state flexible energy storage.

(684f) Activated Carbon and Its Modified Materials From Ddgs' Bio-Char for Supercapacitors

Jin, H., South Dakota State University
Gu, Z., South Dakota State University
Wang, X., South Dakota State University
Originally presented on: 11/7/2013 14:06:00 - 14:22:00

Activated Carbon and its modified Materials from DDGS' Bio-char for Supercapacitors

Super-capacitor is an important electrical energy storage device for promoting widespread application of electrical automobiles, brake energy recovery systems as well as enhancing energy efficient of instruments requiring peak power source, include elevator, crane and locomotives. Pursuing advances in super-capacitors technology is also critical to satisfy needs for developing the future energy storage systems to promote using sustainable wind or solar electricity in rural areas as well as improve application of electrical automobiles. Ultimately, to achieve highly stable reversible electrical energy storage capacity and high power density in supercapacitors, carbon materials to fabricate electrodes, should possess a hierarchical mesoporous structure with high surface, high conductivity, suitable pore size distribution and long-term cycle stability. Until now, the most hopeful research for increasing energy and power density of supercapacitors focused on innovative carbon materials The hierarchical carbon, with high specific surface area (2959 m^2 g^{-1}) and high pore volume (1.65 cm^3 g^{-1}) was prepared from sustainable feedstock - biochar, which was a waste from a thermochemical process optimized for bio-oil production, using KOH catalytic activation. The hierarchical carbon after HNO_3 treatment, as electrode materials, showed improved specific capacitance 260 F g^{-1} in 6 mol L^{-1} KOH at a current density of 0.6A g^{-1}, while the specific capacitance of carbon without oxidation exhibits relative high (200 F g^{-1}) at a higher current density (0.5A g^{-1}) in 6 mol L^{-1} KOH and it can also shows 150 F g^{-1} in the typical organic solution of 1 mol L^{-1} tetraethylammonium tetrafluoroborate in acetonitrile after 1000 cycles. More importantly, the capacitive performances of the hierarchical carbons are much better than general bio-inspired activated carbons, ordered mesoporous carbons and commercial graphene, thus highlighting the success of preparing high performance supercapacitor electrode material from biochar and potential for improving economic viability of thermochemical biofuel processes by converting biochar to a high value added carbon materials.

(684h) Size and Morphology Control of Iron Phosphate- From Nucleation to Growth

Tongbao, Z., Tsinghua University
Yangcheng, L., Tsinghua University
Guangsheng, L., Tsinghua University
Originally presented on: 11/7/2013 14:38:00 - 14:54:00

Iron phosphate is a widely used functional material in the fields of lithium battery, catalysis, adsorption, ferroelectrics and glass and steel industrial. The successful application of iron phosphate in various fields is highly purity, size, morphology and crystal structure dependent. In this work, we developed a common strategy for preparing pure iron phosphate materials with various size and morphology. Composed of fast precipitation in microreactor and following thermal aging treatment, it provided separate and facile process control from nucleation to growth. For examples, as most of iron source precipitated at the nucleation step, size of iron phosphate nanoparticles could be regulated in the range of 9-50 nm by changing supersaturation through coordination effects of inorganic acids; as considerable amount of iron source was delayed for precipitation by simply changing the addition of phosphoric acid, morphology and crystal structure of iron phosphate could be controlled from amorphous nanoparticles to monoclinic microsheets and microspheres in the thermal aging treatment process. Furthermore, we discussed how different mechanisms combined and looked into potential applications with aspects of iron phosphate materials and preparation methods.

AUTHOR INDEX

AUTHOR INDEX

AUTHOR INDEX